Quantum Mec

A Simplified Approach

Quantum Mechanics
A Simplified Approach

Shabnam Siddiqui

CRC Press
Taylor & Francis Group
Boca Raton London New York

CRC Press is an imprint of the
Taylor & Francis Group, an **informa** business

CRC Press
Taylor & Francis Group
6000 Broken Sound Parkway NW, Suite 300
Boca Raton, FL 33487-2742

First issued in paperback 2023

ISBN 13: 978-1-03-265244-3 (pbk)
ISBN 13: 978-1-138-19726-8 (hbk)
ISBN 13: 978-1-315-22315-5 (ebk)

DOI: 10.1201/b22074

This book contains information obtained from authentic and highly regarded sources. Reasonable efforts have been made to publish reliable data and information, but the author and publisher cannot assume responsibility for the validity of all materials or the consequences of their use. The authors and publishers have attempted to trace the copyright holders of all material reproduced in this publication and apologize to copyright holders if permission to publish in this form has not been obtained. If any copyright material has not been acknowledged, please write and let us know so we may rectify in any future reprint.

Publisher's Note
The publisher has gone to great lengths to ensure the quality of this reprint but points out that some imperfections in the original copies may be apparent.

Library of Congress Cataloging-in-Publication Data

Names: Siddiqui, Shabnam, 1975- author.
Title: Quantum mechanics : a simplified approach / Shabnam Siddiqui.
Description: Boca Raton : CRC Press, Taylor & Francis Group, 2018.
Identifiers: LCCN 2018023705 | ISBN 9781138197268 (hardback : alk. paper)
Subjects: LCSH: Quantum theory--Textbooks.
Classification: LCC QC174.12 .S5234 2018 | DDC 530.12--dc23
LC record available at https://lccn.loc.gov/2018023705

Visit the Taylor & Francis Web site at
http://www.taylorandfrancis.com

and the CRC Press Web site at
http://www.crcpress.com

To my parents and family

Contents

Preface xiii
Author xvii
Fundamental Constants and Important Equations xix

Chapter 1   An Introductory History of Quantum Mechanics-I 1

    1.1   Classical View of an Electromagnetic Wave 1
           1.1.1   What Is an Electromagnetic Wave? 2
           1.1.2   What Are the Properties of an Electromagnetic Wave? 3
           1.1.3   How Are the Electromagnetic Waves Generated? 3
    1.2   The Black-Body Radiation Paradox 4
           1.2.1   The Experimental Observations 5
           1.2.2   The Mathematical Form of the Radiation Function 7
    1.3   The Beginning of Quantum Mechanical View 9
    Problems 10
    1.4   The Photoelectric Effect 11
    Problems 14
    1.5   The Compton Effect 14
    Problems 16
    1.6   Tutorial 17
           1.6.1   Photoelectric Effect 17

Chapter 2   An Introductory History of Quantum Mechanics-II 21

    2.1   De Broglie Matter Waves 21
    Problems 25
    2.2   Electron Diffraction 25
    Problems 27
    2.3   Double-Slit Experiment 27
    Problems 32
    2.4   Uncertainty Principle 32
           2.4.1   The Uncertainty Principle 33
           2.4.2   A Simple Proof of the Uncertainty Principle 35
    2.5   Measurement 36

Chapter 3   Formalism 39

    3.1   Introduction 39
    3.2   Describing a Quantum Mechanical System: Basic Postulates of
           the Model 39
           3.2.1   Postulate 1: Defining an Observable 39
           3.2.2   Indeterminacy Associated with the Measurement of an
                     Observable 40

      3.2.3   Postulate 2: Defining the States Associated with
                 an Observable 40
      3.2.4   Postulate 3: Measurement of an Observable 41
3.3   Mathematical Foundation 41
      3.3.1   State Vector 41
      3.3.2   Ket and Bra Space 42
      3.3.3   Multiplication Rules for Ket and Bra.......................... 43
           3.3.3.1   Inner Product 43
           3.3.3.2   Outer Product...................................... 43
           3.3.3.3   Orthogonality 43
           3.3.3.4   Normalization..................................... 43
      3.3.4   Operators... 45
      3.3.5   Base Kets 45
      3.3.6   Completeness.. 45
           3.3.6.1   Discrete Spectrum.................................. 48
           3.3.6.2   Continuous Spectrum 50
Problems.. 50
3.4   Wave Function 51
      3.4.1   Function Space and Hilbert Space 51
      3.4.2   Eigenfunctions.. 53
           3.4.2.1   Orthogonal 53
           3.4.2.2   Normal 54
           3.4.2.3   Completeness...................................... 54
      3.4.3   Probability Density 54
      3.4.4   Expectation Value of an Observable 56
           3.4.4.1   Position 56
           3.4.4.2   Momentum 57
           3.4.4.3   Kinetic Energy..................................... 57
Problems.. 58
3.5   Dynamical Properties of the System 59
      3.5.1   Time-Dependent Schrodinger Equation 59
      3.5.2   Time Evolution Operator...................................... 60
      3.5.3   Separation of Variables: Time-Independent
            Schrodinger Equation 61
      3.5.4   Stationary States 62
Problems.. 63
3.6   Conservation of Probability 64
Problem 65
3.7   Heisenberg Uncertainty Principle 65
      3.7.1   Proof of the Uncertainty Principle 67
           3.7.1.1   The Schwarz Inequality 67
      3.7.2   Applications of the Principle 68
           3.7.2.1   Heisenberg's Microscope........................... 68
           3.7.2.2   Defining Orbits in Atoms........................... 69
Problems.. 70
3.8   Matrix Mechanics 71
3.9   Tutorials 73
      3.9.1   Spin-1/2 73
      3.9.2   Wave Function 75

Chapter 4 Applications of the Formalism-I... 81

  4.1   Introduction 81
  4.2   The Free Particle 81
  Problems... 86
  4.3   The Infinite Square Well 86
  Problems... 92
  4.4   Step Potential... 92
  Problems... 98
  4.5   Potential Barrier Penetration (Tunneling)............................ 98
  Problems.. 100
  4.6   The Finite Square Well 100
  Problems.. 104
  4.7   Tutorials 105
        4.7.1   Infinite Square Well.. 105
        4.7.2   Tunneling 110
        4.7.3   Quantum Wave Packet 119

Chapter 5 Applications of the Formalism-II.. 127

  5.1   The Harmonic Oscillator 127
        5.1.1   Analytical Method 127
                5.1.1.1   Normalization Constant 131
  Problems.. 133
        5.1.2   Algebraic Method... 134
                5.1.2.1   Normalization Constant 137
  Problems.. 138
  5.2   The Schrodinger Equation in Three Dimensions........................ 138
        5.2.1   The Schrodinger Equation in Cartesian Coordinates 138
                5.2.1.1   What Is the New Quantum Property of the
                          3-Dimensional Infinite Square Well? 140
  Problems.. 141
        5.2.2   The Schrodinger Equation in Spherical Coordinates 141
        5.2.3   The Angular Equation....................................... 142
        5.2.4   The Radial Equation 145
  5.3   The Hydrogen Atom... 146
        5.3.1   The Radial Equation for the Hydrogen Atom.................. 147
                5.3.1.1   Asymptotic Behavior.............................. 147
                5.3.1.2   Power Series Solution 148
                5.3.1.3   The Laguerre Polynomials and the Associated
                          Laguerre Polynomials 151
                5.3.1.4   Degeneracy of Hydrogen Atom 153
        5.3.2   Hydrogen Atom Spectrum..................................... 153
  Problems.. 156
  5.4   The Angular Momentum 156
        5.4.1   What Are the Eigenfunctions and Eigenvalues
                of These Operators?.. 159
        5.4.2   Is Angular Momentum Conserved in a Quantum
                Mechanical System? 159

Problems.. 161
5.5   Tutorial 162
        5.5.1   Angular Momentum 162

Chapter 6   Perturbation Theory 167

6.1   Time-Independent Perturbation..................................... 167
        6.1.1   Non-Degenerate Perturbation 168
                  6.1.1.1   First-Order Correction........................... 168
                  6.1.1.2   Second-Order Correction 170
        6.1.2   Degenerate Perturbation................................. 170
                  6.1.2.1   What Is a Degenerate State? 170
        6.1.3   Stark Effect.. 172
                  6.1.3.1   Parity... 172
Problems.. 173
6.2   The Variational Principle.. 174
        6.2.1   The Ground State of a Helium Atom......................... 176
Problems.. 180
6.3   The WKB Method.. 180
        6.3.1   Turning Points of a Bound State 186
Problems.. 188

Chapter 7   Time-Dependent Perturbation Theory........................ 189

7.1   Introduction 189
7.2   First-Order Perturbation Theory 190
7.3   Periodic Perturbations... 193
Problems.. 194
7.4   The Sudden Approximation 194
Problems.. 196
7.5   Adiabatic Approximation 196
        7.5.1   Adiabatic Process... 197
        7.5.2   Adiabatic Theorem... 197
        7.5.3   Proof of the Adiabatic Theorem 199
Problems.. 202
7.6   Measurement Problem Revisited..................................... 202
        7.6.1   EPR Paradox 204
        7.6.2   Bell's Inequality 206

Chapter 8   Quantum Computer.. 209

8.1   Classical Computer 209
8.2   Quantum Computer.. 211
        8.2.1   Qubit... 212
        8.2.2   Multiple Qubits... 214
                  8.2.2.1   Entanglement 215
        8.2.3   Qubit Gates... 215
Problems.. 217
8.3   Quantum Algorithms.. 218

8.4   Quantum Measurement 220
8.5   Density Operator 222
        8.5.1   Properties of a Density Operator 223
        8.5.2   Reduced Density Operator................................... 224
Problems... 224
8.6   Decoherence 224
8.7   Methods for Overcoming Decoherence 226
        8.7.1   Quantum Error-Correcting Codes 226
Problems... 228
8.8   Quantum Teleportation... 228
8.9   Superdense Coding 230

References 231
Index 233

Preface

The quantum theory has proven to be a very successful theory to describe the fundamental reality of nature. Its predictions for fundamental natural phenomenon (e.g., atomic structure, radiation and band spectra, etc.) have proven to be accurate in numerous and varied experimental conditions. Such explanations cannot be derived from classical physics. However, the development of quantum theory has led to a radical shift in the understanding of the physical universe that was derived from classical physics. Our day-to-day reality is based on intuition developed from the concepts of classical physics. This radical departure from our deep thought processes seeded in classical reality challenges our ability to realign our thinking to accommodate the paradigm shift of quantum mechanics. One possibility to reprogram our instincts to properly grasp this paradigm shift is to proceed with small steps in the study of individual phenomenon that manifest the differences between the two systems of thought, for example, the double-slit experiment, quantum tunneling and the Stern-Gerlach experiment. An in-depth study of each case should include a direct comparison and contrast between the predictions of quantum mechanics with those of classical theory. This same approach is followed in this book at a relatively elementary level. Every chapter presents quantum ideas step-by-step in a structured way, with a comparison between quantum and classical ideas to assist the reader in grasping the fundamental differences between them.

The main goal of this book is, therefore, to understand quantum mechanics through the process of conceptual questioning and problem solving in this new paradigm. Problem solving in this context is not restricted to solving differential equations and integration, but rather it involves systematical thinking about the problem to apply the new and powerful concepts of quantum mechanics to find a solution and understand the physical meaning of that solution. Each topic of quantum mechanics is introduced here by applying a step-by-step approach to facilitate the learning process. The systematic arrangement of the topics is meant to assist the reader in the construction of a sound foundation in the field. For example, the first topic discussed is that of a quantum free particle and subsequently that of a quantum particle bound by a potential. Understanding the dynamics and the potential of a free particle is a critical first step that is required before tackling the more complex case of a bound particle. Reversing this order would create confusion for the reader, and such a systematic approach for the arrangement of the topics has been applied for all chapters of the book.

The tutorials on special topics at the end of chapter are an effort to teach problem solving by actively engaging the reader in a thinking process about the topic in quantum mechanical theory. This process should assist the reader toward the development of a new "quantum mechanical intuition" in understanding nature. For example, in one of the tutorials on quantum tunneling, the concept of a classical particle and a plane wave approaching a barrier is introduced, and the energy conditions for the system that are required to overcome the barrier are interrogated. After exploring the classical concept, a quantum mechanical particle approaching a barrier is also explored, and the differences between the two cases are investigated in some detail. Since a quantum particle is both a wave and a particle, it behaves in a completely different manner than the classical particle or wave. By asking step-by-step questions concerning the differences between the two cases, such as What are the conditions that allow a quantum particle to tunnel through the barrier? and What are the conditions that cause a quantum particle to reflect from the boundary? the reader is encouraged to consider the tunneling process in depth by asking the right questions. They are thus primed to engage in a deep-thinking process about the details of the quantum mechanical effect for each case.

Some of the tutorials include links to simulations. These simulations are helpful in a visualization of quantum phenomenon and at the same time aid in an understanding of the application of the mathematics. The tutorials can be used by a teacher as take-home exercises, projects, and in-class group exercises. In addition to the tutorials, the book includes conceptual questions placed appropriately in between the sections of the chapters. The conceptual questions are introduced to test a logical understanding of the concepts. The problems at the end each section provide additional practice in applying concepts and scientific reasoning skills. This book is an attempt to offer a combination of traditional and nontraditional approaches to learning. The key features of this book are:

- A simplified, structured and step-by-step introduction to quantum mechanics at an elementary level
- A systematic arrangement of topics and chapters
- Tutorials on special topics to actively engage students in a thinking process concerning the differences between quantum mechanics and classical physics
- Conceptual questions to test logical understanding
- Traditional problems at the end of each section for additional practice
- Simulations to aid in the visualization of the physical phenomenon and a demonstration of the application of mathematics

In this book, the first two chapters are dedicated to a historical introduction to quantum mechanics. Such an introduction is necessary to build a sound foundation of the important experiments and ideas that lead to the early formulation of the theory of quantum mechanics. As a first step toward the development of a mathematical foundation of the theory, a formalism of quantum mechanics is discussed in some detail in Chapter 3. The formalism of quantum mechanics is one of the most abstract and difficult topics to learn. This chapter is presented in a very structured way, starting with an introduction of the linear superposition state and the need for such a "state" as a description of a physical system in quantum mechanics. First, the concept of a state vector is introduced, and then the derivation of a wave function from the state vector is shown. State vectors represent superposed discrete states that are a key "quantum feature," and if introduced only through the wave function, which is a function of a continuous variable, it may lead to a gap in understanding of this key building block of the theory. Later, a solution of the Schrodinger equation for a given wave function is discussed. The physical meaning of the mathematics is explained to improve further an understanding of quantum phenomenon for most of the topics in this chapter. In this book, for all chapters, Schrodinger's "wave mechanics" approach is applied. Heisenberg's "matrix mechanics" is not discussed in order to reduce confusion for the reader in applying the theoretical foundations of the theory. However, a brief introduction to matrix mechanics is offered in Chapter 3. In Chapters 4 and 5, the applications of formalism for several quantum systems such as a free particle, an infinite square well, finite potentials, a harmonic oscillator, the hydrogen atom and the atomic spectrum of hydrogen are discussed at some length. These applications are focused toward the development of a quantum mechanical intuition of these illustrative quantum mechanical phenomenon. The emphasis is on an understanding of the physical system, solving the Schrodinger equation for the system, and an interpretation of the physical phenomenon exhibited by the system obtained from the solution of the Schrodinger equation. The main purpose, therefore, is to offer the reader a solid understanding of the mathematical techniques that are typically applied for solving the Schrodinger equation, along with a relatively complete understanding of the quantum behavior of the system. Mathematical functions such as the Fourier transform, the Legendre polynomial and Hermite polynomials are presented in a "casual manner" to induce a sense of familiarity with the reader for these concepts. These are interesting mathematical functions with special features that are useful in an understanding of the fundamentals of quantum theory. These special features are the focus of the introduction of these functions. In later chapters, time-independent and time-dependent perturbation theories are discussed. The focus of these chapters is to demonstrate that for some systems the Schrodinger equation cannot be solved exactly, and therefore approximate methods can be applied. These chapters will prepare students for more practical applications of quantum mechanics. In the latter part of Chapter 7, the problem of measurement is discussed comprehensively. The EPR paradox and Bell's inequality are also discussed here. This section offers a broader view of the problem and how it was solved. In the last chapter, the quantum computer is discussed as an important emerging application of quantum mechanics.

Since the quantum computer will soon be a reality, it is important to introduce such powerful and practical concepts to students. The book begins with Planck's idea of "quanta" and finishes with a solid introduction to the quantum computer. The book is intended for a one-semester or one-year course in quantum mechanics at the junior or senior level.

I have many people to thank, as without their support, this work would have never become a reality. I would like to acknowledge my colleagues, and mentors who have helped me with their suggestions and advice for improving the chapters from the book. I am grateful to Lee Sawyer, Julio Gea-Banacloche, B. Ramachandran and Leon D. Iasemidis. For help with the figures and index, I would like to thank Haocheng Yin and Imran Hossain. I would also like to thank Ian Wylie for help with editing, and the CRC Press team, Marc Gutierrez, Judith Simon and Ragesh K. Nair, for making it easy for me, and Prabhu Arumugam for his constant support.

Author

Dr. Shabnam Siddiqui is a research assistant professor at Louisiana Tech University, LA. She teaches physics and conducts research for developing electrochemical microsensors using carbon nanomaterials. She applies active learning approaches for teaching physics courses and focuses on developing new methods for learning physics, and quantum mechanics. Dr. Siddiqui studies properties of carbon nanomaterials for attaining reliable and real-time sensing. She has authored over 20 peer reviewed journal papers. She earned a PhD in physics in quantum computing and quantum information in 2006 from the University of Arkansas at Fayetteville, AR. Dr. Siddiqui received post-doctoral training at NASA Ames Research Center, CA, and the University of Pittsburgh, PA. She had also worked at Advanced Diamond Technologies, IL prior to joining Louisiana Tech.

Fundamental Constants and Important Equations

FUNDAMENTAL CONSTANTS

Planck's constant: $\hbar = 1.05457 \times 10^{-34}$ Js

Boltzmann constant: $k_B = 1.38065 \times 10^{-23}$ J/K

Speed of light: $c = 2.9979 \times 10^8$ m/s

Charge of electron: $-e = -1.60218 \times 10^{-19}$ C

Charge of proton: $+e = +1.60218 \times 10^{-19}$ C

Mass of electron: $m_e = 9.10938 \times 10^{-31}$ kg

Mass of proton: $m_p = 1.67262 \times 10^{-27}$ kg

Permittivity of space: $\varepsilon_0 = 8.85419 \times 10^{-12}$ C^2/Jm

Bohr Radius: $a = \dfrac{4\pi\varepsilon_0\hbar^2}{me^2} = 5.29177 \times 10^{-11}$ m

Coulomb constant: $\dfrac{1}{4\pi\varepsilon_0} = 8.98755 \times 10^9$ N.m^2/C^2

USEFUL INTEGRALS

$$\int \sin\left(\frac{n\pi x}{L}\right) dx = \frac{x}{2} - \frac{L}{4n\pi}\sin\left(\frac{2n\pi x}{L}\right)$$

$$\int x \sin\left(\frac{n\pi x}{L}\right) dx = \frac{x^2}{4} - \frac{Lx}{4n\pi}\sin\left(\frac{2n\pi x}{L}\right) - \frac{L^2}{8n^2\pi^2}\cos\left(\frac{2n\pi x}{L}\right)$$

$$\int_0^\infty x^n e^{-bx} dx = \frac{n!}{b^{n+1}}$$

$$\int_{-\infty}^\infty e^{-x^2} dx = \sqrt{\pi}$$

$$\int_{-\infty}^\infty e^{-a(x-b)^2} dx = \sqrt{\frac{\pi}{a}}$$

$$\int_{-\infty}^\infty x^2 e^{-bx^2} dx = \frac{1}{2}\sqrt{\frac{\pi}{b^3}}$$

1 An Introductory History of Quantum Mechanics-I

1.1 CLASSICAL VIEW OF AN ELECTROMAGNETIC WAVE

The origin of quantum mechanics is linked to the discovery of then new laws of radiation by Max Planck in 1900. The explanation of the empirical facts of radiation by Planck using an extraordinarily powerful yet simple model lead to the discovery of the "quantum" nature of radiation. According to his model, radiation is emitted or absorbed in "bits" or packets of discrete energy. His formula for the intensity of radiation modeled the experimental results very well. His radiation model turned out to be a much closer simulation of the experimental results. However, this finding that radiation is emitted or absorbed in discrete "quanta" rather continuous energy flow was profoundly novel idea unexpected to classical physicists at that time. This new idea disturbed the foundations of the classical theory of radiation. At first, even Planck was skeptical about his own idea and spent time and effort in attempting to reconcile his model with the classical theory. However, he failed in his attempt, since there was no realistic means for him to reconcile quantum theory with the classical theory. Therefore, he published his novel theory in 1900. It took another five years for Planck's new theory to be applied to the solution of a concrete physical problem. It was Albert Einstein in 1905 who extended Planck's idea to an explanation of the specific heats of solids and of gases at low temperature and to the photoelectric effect. In both cases, Einstein was able to demonstrate that by applying Planck's model of the quantization of energy, these phenomena can be readily explained in detail. However, this also marked the beginning of the understanding of light as a phenomenon with a discrete nature, which was unknown to classical physics at that time. Einstein was able to explain the photoelectric effect using Planck's idea of quantization of energy and showed that light can be discrete as well. According to Maxwell's theory, light is an electromagnetic wave, and thus is continuous. Well-known phenomenon such as diffraction and interference could only be explained by considering light as a continuous system such as an electromagnetic wave. This new idea about the discrete nature of light (consisting of particles) turned out to be contradictory to the Maxwell's theory of light. But could light have a dual nature? Can it be both discrete and continuous? How can something be discrete and continuous at the same time? Similar paradoxes also arose in atomic physics. In 1913, Niels Bohr applied these new ideas to atomic physics by explaining the spectrum of the hydrogen atom. However, it turned out that there was a discrepancy between the calculated orbital frequency of the electrons and the frequency of the emitted radiation. Further, in 1924, Compton discovered that the frequency of X-rays changes as they collide with electrons. This could only be explained by considering X-rays as discrete particles colliding with electrons. All these experimental observations and the failure of classical physics to explain these new findings related to the atomic phenomenon paved the way for the development of a mathematical theory of quantum mechanics. The work to develop a mathematical theory of quantum mechanics continued for a very long time (over 30 years). Finally, two mathematical approaches for the formalism of quantum mechanics emerged. One was developed in 1925 by Werner Heisenberg, called matrix mechanics, and the other was developed by Erwin Schrodinger in 1926 called wave mechanics. The mathematical formalism of both approaches turned out to be equivalent. Applying the mathematical theory of quantum mechanics allowed all quantum paradoxes described previously to be explained, and this led to its application in the fields of optics, thermodynamics, chemistry and mechanics. Quantum mechanics has been very successful in providing an understanding of the fundamental nature of matter and light. This understanding has led to the development of devices such as lasers, diodes, transistors, electron microscope, magnetic resonance imaging and many more. These devices are the basis of today's electronics, computers and telecommunications. In this way, our modern life has a strong connection to the remarkable discovery of the quantum theory of radiation by Planck. Without Planck's discovery, it is difficult to imagine how these devices would have come to exist or at least to be discovered on the same timeline.

It seems that our curiosity about light and matter has led us to discover the unknown truths about it. The more we learned about it, the more it has allowed us to know the truths about our environment, universe and about ourselves. The story of quantum mechanics began with discoveries of the properties of radiation, light and matter, and yet, the picture is not complete. To date, quantum mechanics has been applied to the fields of physics, chemistry, biology, information and computation. A full understanding of the fundamental quantum behavior of information and life in these fields is still very much in progress. As of now, it is difficult to predict what will transpire in these emerging applications of quantum mechanics even in the near future. However, it is equally clear that efforts in these fields will produce worthwhile and even extraordinary benefits.

After this brief introduction on the historical basis of quantum theory, we can begin here by understanding the classical view of an electromagnetic wave and proceed step-by-step in exploring the historical development of quantum mechanics. We will not be covering each discovery that led to the development of quantum mechanics but rather the most relevant to the current understanding of the theory.

1.1.1 What Is an Electromagnetic Wave?

The waves that we see around us such as ripples in a pond, and oscillations of a rope tied to a support can be defined simply as a disturbance that propagates through a continuous medium at a constant velocity. However, all waves do not require a medium (continuous or discrete) to propagate. There are waves such as electromagnetic waves that propagate without a medium.

In classical physics, an electromagnetic wave is defined as oscillating electric and magnetic fields in a plane (oscillating fields as disturbance) that are perpendicular to the direction of propagation of the wave. Such waves move with the speed of light in a vacuum. Light and any form of radiation is an electromagnetic wave. The Figure 1.1 illustrates electromagnetic waves and ripples in a pond.

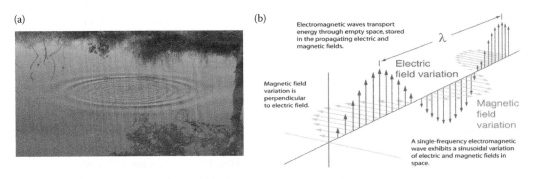

FIGURE 1.1 (a) This figure illustrates ripples in a pond as waves that are disturbance in a continuous medium that is water (Credit: Kris Bridgman). (b) This figure illustrates electromagnetic waves as oscillating electric and magnetic fields. Such waves do not require a medium such as water to propagate. (Taken from HyperPhysics website, Credit: Dr. Rod Nave)

1.1.2 WHAT ARE THE PROPERTIES OF AN ELECTROMAGNETIC WAVE?

The electric and magnetic fields associated with electromagnetic waves carry energy. Such waves transport energy as they move in space. When electromagnetic waves impinge upon matter, they can transfer energy to it. This can lead to ionization of material due to breaking of the molecular bonds. Also, such waves can travel very long distances, and they can be reflected and absorbed by any material object in their path. There are many types of electromagnetic waves. However, such waves can be distinguished from each other by their frequency and wavelength. The frequency of an electromagnetic wave is the number of oscillations per second, wavelength is the distance at any instant between two consecutive peaks or valleys, and amplitude is the distance from the center line to the top of the peak or valley. The frequency and wavelength of such waves are inversely proportional to each other. The waves that have high frequency will be characterized by a small wavelength and vice versa. In a vacuum, the frequency of an electromagnetic wave is expressed by the following relationship:

$$\nu = \frac{c}{\lambda} \tag{1.1}$$

where c is the speed of light (3×10^8 m/s), ν is frequency and λ is wavelength.

Thus, all electromagnetic waves, no matter what their frequency is, travel with the speed of light in a vacuum.

1.1.3 HOW ARE THE ELECTROMAGNETIC WAVES GENERATED?

An accelerating or oscillating electric charge such as an electron generates an oscillating electric field, and an oscillating electric field generates an oscillating magnetic field. These fields are perpendicular

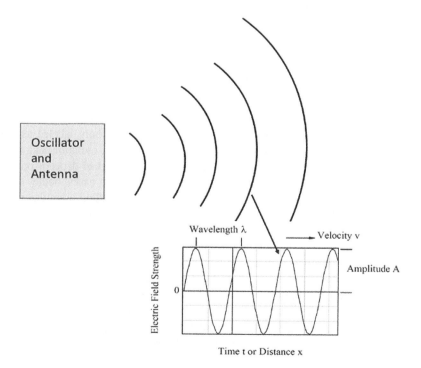

FIGURE 1.2 An electronic oscillator connected to an antenna produces electromagnetic waves that propagate in space.

to each other and produce electromagnetic waves that travel with some velocity. The frequency of the electromagnetic wave is the same as the frequency of mechanical oscillation of the charge, and the intensity of the electromagnetic wave is the square of the amplitude of the mechanical oscillation of the charge. An electronic oscillator connected to an antenna produces electromagnetic waves. By constructing an oscillator that oscillates at any frequency, different electromagnetic waves with different frequencies can be generated. The schematic in Figure 1.2 shows generation of electromagnetic waves by an oscillator.

The electromagnetic spectrum shown in Figure 1.3 shows all types of electromagnetic waves.

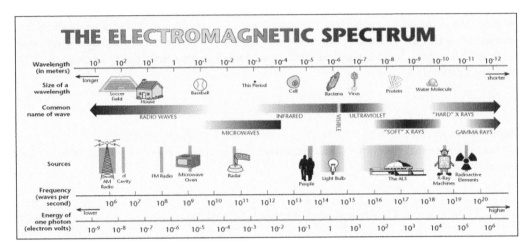

FIGURE 1.3   The electromagnetic spectrum covers a wide range of wavelengths and photon energies. Light used to see an object must have a wavelength about the same size or smaller than the object. This image is taken from NASA website, https://www.nasa.gov/centers/glenn/about/fs13grc.html, with permission under the public domain license.

1.2   THE BLACK-BODY RADIATION PARADOX

Radiation is defined as an electromagnetic wave, and it can be any type of electromagnetic wave, as shown in the previous spectrum. Radio waves, microwaves, X-rays and gamma waves are part of the same general "type" of electromagnetic waves that are commonly labeled as "radiation." All materials or bodies that have inherent temperature $(T \neq 0 \text{ K})$ emit radiation and absorb radiation when it falls upon them. The amount of radiation emitted or absorbed depends on the material properties of that body. In general, all bodies fall between two limiting cases, bodies that are "perfectly reflecting," such that they completely reflect all the radiations falling upon them and emit no radiations when heated. The opposite case of a perfectly reflecting body is called "black body," and such a body absorbs all the radiations falling upon it and emits all the radiations when heated. In fact, there are no materials that exhibit such properties, but there are materials whose behavior can be closely approximated by a black or perfectly reflecting body. For example, "polished silver" can be approximated as perfectly reflecting body, and "soot" can be approximated as a black body. The emission of radiation by black bodies was one of the least understood phenomenon during the late nineteenth century, and attempts were made by physicists at that time to develop a theory of black-body radiation. Interestingly, the development of the theory of black-body radiation presented utmost challenges to the physicists at that time. In fact, new ideas were sought to explain the phenomenon of black-body radiation, since classical theories turned out be insufficient to explain such a familiar phenomenon. These new ideas laid the foundation of the quantum theory.

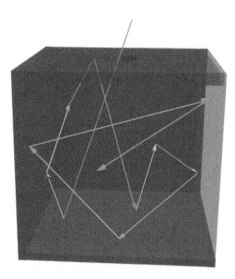

FIGURE 1.4 A schematic of a black body. The radiation inside the enclosure is perfectly reflected from all the walls.

1.2.1 THE EXPERIMENTAL OBSERVATIONS

The black bodies available naturally are not perfect black bodies. However, radiation inside a heated closed enclosure kept at a constant temperature can be considered as a perfect black body (Figure 1.4). Therefore, experiments were performed using such black bodies to understand the properties of radiation. The radiation inside such an enclosure behaves very similar to gas. It exerts pressure against the walls of the enclosure, work must be done to compress it to a smaller volume, and if the enclosure remains closed, the characteristics of the radiation remains unchanged even after a very long period of time. Thus, such radiation is named radiation gas (Figure 1.4).

Inside the enclosure, it is only radiation gas and no matter, and walls of the enclosure are perfectly reflecting. Hence, the radiation has no source inside the enclosure for its absorption and emission. Suppose that inside such an enclosure maintained at constant temperature a piece of matter is introduced into the box such as any approximate black body such as "soot." This soot is capable of absorbing radiations of all frequencies and at the same time emits radiations of all frequencies. The absorption and emission will continue until an equilibrium is reached, and the radiation inside the enclosure remains unchanged in its composition. Such a state of radiation is called as a state of statistical equilibrium, and the name equilibrium radiation is given to a mixture of radiation of different frequencies that will be found in the enclosure.

Several experiments were performed using such a black body, and several properties of the radiation were derived, in the form of the following laws.

1. *Kirchhoff's law*: According to this law, for a given temperature, the composition of the equilibrium radiation inside the enclosure is exactly the same regardless of the nature of matter. The intensity of the radiation coming out of the enclosure is independent of the size and shape of the enclosure. The intensity of the monochromatic radiation is solely dependent on the frequency of the radiation and the temperature of the enclosure. Therefore, intensity can be represented as a function of frequency and temperature

$$I_\nu = I(\nu, T) \tag{1.2}$$

The intensity of the monochromatic radiation is generally related to the differential amount of radiant energy, dE, that crosses an area element dA, in directions confined to a differential solid angle $d\Omega$ being oriented at an angle θ to the normal of dA in the time interval t and $t + dt$ and the frequency interval ν and $\nu + d\nu$.

Thus, the differential radiant energy in the form of intensity function can be written as:

$$dE = I(\nu, T)\cos\theta \; dA \; d\Omega \; d\nu \, dt \tag{1.3}$$

2. *Stefan-Boltzmann's law*: According to this law, the total energy radiated per unit surface area of the black body per unit time for all frequencies of the radiation is proportional to the fourth power of the absolute temperature "T" of the enclosure.

$$J = \varepsilon\sigma \; T^4 \tag{1.4}$$

ε = emissivity, σ = Stefan-Boltzmann constant, J = the energy per unit area of the black body and per unit time, and T = temperature in Kelvin.

3. *Wein's Displacement Law*: The direct measurements exhibited in the graph of Figure 1.5 illustrate how the intensity of radiation varies with frequency for different temperatures. As the temperature increases, so does the intensity of the radiation. The range of frequency of radiation emitted for any given temperature is continuous with a specific wavelength of radiation more intense than the others. The frequency of the highest intensity radiation is directly proportional to the absolute temperature. This is called the Wein's displacement law. Equation (1.5) describes this law:

$$\nu_{\max} = C\,T; \quad \lambda_{\max}T = 2.898 \times 10^{-3}\,\text{m.K} \tag{1.5}$$

where C = a constant, and T = absolute temperature in Kelvins.

All the previously mentioned laws are empirical laws and provide a sound basis for developing a theoretical model for explaining the phenomenon of radiation. These empirical laws obtained directly from experimental measurements laid the foundation for developing general laws of radiation.

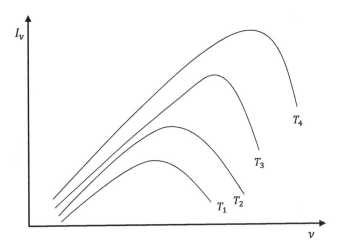

FIGURE 1.5   The intensity Vs frequency graph of radiation obtained from the experimental observations. The different lines show the variation of the intensity with frequency for different temperatures. The higher the temperature of the enclosure, the higher is the intensity of radiation of maximum frequency.

1.2.2 THE MATHEMATICAL FORM OF THE RADIATION FUNCTION

The quest for more general laws that describe mathematically the form of the intensity function (Equation 1.2) as discussed previously lead to the following findings.

1. *Wein's Relation*: Before Wein, the mathematical form of the intensity function for radiation was not known. The form of the function described by him was of special importance. According to Wein, the intensity of the radiation of frequency ν inside an enclosure at temperature T can be described by the mathematical equation:

$$I_\nu = \frac{\nu^3}{c^2} \ F\left(\frac{\nu}{T}\right) \tag{1.6}$$

where $F\left(\frac{\nu}{T}\right) =$ an unknown radiation function, $c =$ the velocity of light in a vacuum, and $\nu =$ the frequency of radiation. Before Wein, the form of the radiation function was unknown. According to Wein, the function $F\left(\frac{\nu}{T}\right)$ has the following form:

$$F\left(\frac{\nu}{T}\right) = e^{-\frac{\nu}{T}} \tag{1.7}$$

This development brought a new insight into formulating the mathematical form of the radiation function.

2. *Rayleigh–Jeans law, the black-body paradox*: Rayleigh and Jeans designed the mathematical form of the previous radiation function based on the classical laws of electromagnetism and thermodynamics. They obtained the following relationship purely based on the classical laws:

$$I_\nu = \frac{\nu^2}{c^2} \ k\,T \tag{1.8}$$

$k =$ Boltzmann constant, and $T =$ temperature in Kelvin.

The differential energy of radiation of frequency range ν and $\nu + d\nu$ within a volume dV is given as:

$$dE = \frac{\nu^2}{c^2}\, kT\, dV\, d\nu \tag{1.9}$$

The above relationships indicates that radiation of higher frequencies such as ultraviolet light will have higher intensity. Therefore, according to this equation, the intensity of radiation of infinite frequency should be most intense and the intensity versus frequency graph shown in Figure 1.5 will diverge at higher frequencies. By integrating Equation 1.9 over frequency range from 0 to ∞ leads to infinite energy of the radiation inside an enclosure at temperature "T." This conclusion completely contradicted experimental findings. It is also incompatible with Stefan-Boltzmann's law of radiation. This is called the "ultraviolet catastrophe or black body paradox." This finding by Ryleigh and Jeans was important, as it clearly demonstrated that the classical view of the radiation as an electromagnetic wave violated the experimental observations.

3. *Planck's radiation model*: According to classical physics, the equilibrium radiation is established inside the enclosure when the emission and absorption of radiation by matter occurs at the same rate. This matter can be assumed to be comprised of electrons that can be represented as oscillators that vibrate back and forth along a straight line. Each oscillator can emit or absorb radiation of a certain frequency. This frequency is called the "characteristic

frequency of vibration" of each oscillator. A state of equilibrium is achieved between the oscillators and the radiation inside the enclosure. The emission and absorption of radiation by the oscillators leads to fluctuations in their energy values. Therefore, oscillators of the same characteristic frequency will exhibit different energies at any given instant. The average energy of all the oscillators, whose characteristic frequency is "ν" and in a state of statistical equilibrium at a temperature "T" of the enclosure, can be represented by a function $U(\nu, T)$. The quantity $U(\nu, T)$ is called the "monochromatic" energy density or the "spectral energy density" of radiation. Using this function, the total intensity of radiation of frequency "ν" inside the enclosure can be written as:

$$I(\nu, T) = \frac{c}{4\pi} U(\nu, T) \tag{1.10}$$

The continuous emission and absorption of such radiation by the oscillators lead to the following form of the function:

$$U(\nu, T) = \frac{8\pi\nu^2}{c^3} k T \tag{1.11}$$

$k = $ Boltzmann constant, and $T = $ the absolute temperature of the enclosure.

However, the previous discussion clearly demonstrates that such a form of the energy function of the oscillator lead to erroneous law of Rayleigh–Jeans. Planck was guided by these difficulties to devise a new theory for explaining the experimental observations of the radiation. He came up with a new theory, named "quantum theory." In this theory, he postulated that emission and absorption of radiation is a discontinuous process. Thus, the total energy of oscillator of frequency "ν," can only have certain values:

$$0, \quad \nu, \quad 2h\nu, \quad 3h\nu, \quad 4h\nu, \quad 5\ h\nu, \ldots nh\ \nu$$

and energy is radiated and absorbed only as packets of energy $h\nu$. Here, h is called as Planck's constant, and its value is equal to 6.626×10^{-34} J s.

These assumptions led to the following equation for the energy density function of an oscillator:

$$U(\nu, T) = \frac{8\pi\nu^2}{c^3} \frac{h\nu}{\left(e^{\frac{h\nu}{kT}} - 1\right)} \tag{1.12}$$

According to Planck, this function may be related to the average energy ($\bar{E}$) of the oscillator of frequency ν inside the enclosure. Such a function satisfies all three of the previous radiation laws and was in agreement with experimental measurements at all temperatures. Using this function, the intensity function of the radiation can be written as:

$$I(\nu, T) = \frac{2h\nu^3}{c^2} \frac{1}{\left(e^{\frac{h\nu}{kT}} - 1\right)} \tag{1.13}$$

In terms of the wavelength, the previous function can also be written in the following form, We will use the following relations to derive the wavelength dependent intensity function:

$$\frac{c}{\lambda} = \nu$$

Thus,

$$\frac{-c}{\lambda^2} d\lambda = d\nu \tag{1.14}$$

The differential energy of radiation of frequency range ν and $\nu + d\nu$, in volume dV, in time $t, t + dt$, is given as,

$$dE = I(\nu, T) \cos\theta \ dA \ d\Omega \ d\nu \ dt$$

where $d\Omega$ is the differential solid angle, being oriented at angle θ to the normal of area dA.

In terms of wavelength, the previous equation can be written as:

$$dE = I(\lambda, T) \cos\theta \ dA \ d\Omega \ dt \left(\frac{-c}{\lambda^2}\right) d\lambda \tag{1.15}$$

Therefore, intensity function in terms of wavelength is:

$$I(\lambda, T) = -\frac{2h \ c^2}{\lambda^5} \frac{1}{\left(e^{\frac{h\nu}{kT}} - 1\right)} \tag{1.16}$$

According to Planck, it is the energy of the radiation that is emitted or absorbed as packets of energy called "quanta." Therefore, it is the energy of radiation that is quantized. It cannot take any value as postulated by classical theory.

1.3 THE BEGINNING OF QUANTUM MECHANICAL VIEW

The pioneering work of Planck led to the new law of radiation known as Planck's radiation law. This new law can be summarized as following:

Planck's radiation law: The sources of thermal radiation are atoms in a state of oscillations, and vibrational energy of each oscillator may have any of a series of discrete values but never any value in between. The energy E an oscillating atom can have is given as:

$$E_n = n \ h \ \nu \tag{1.17}$$

where $n = 1, 2, 3, \ldots$, $h =$ Planck's constant, and $\nu =$ the frequency of the oscillator.

When an oscillator (oscillating atom) changes from a higher state of energy E_2 to a state of lower energy E_1, it emits discrete amount of energy as $E_2 - E_1 = h \ \nu$. This smallest amount of energy is called as *quanta of energy*. When an oscillator changes its energy state from lower (E_1) to a higher energy state (E_2), it absorbs a quantum of energy. Planck illustrated the mechanism of emission and absorption of radiation and the fact that it is only the energy of the radiation that is quantized. He still considered the radiation to be an electromagnetic wave.

Later, this idea was further advanced by Einstein, who suggested that light, which is an electromagnetic wave, is made up of discrete elementary particles called photons. The energy carried by a photon is a function of its frequency: $E_{photon} = h\nu$. We will discuss this in more details in the next section. The schematic that follows describes the change in views of radiation after the discovery of Planck (Figure 1.6).

Conceptual Question 1: Explain in your own words the differences between classical and quantum mechanical view of radiation.

Classical View   **Quantum mechanical View**

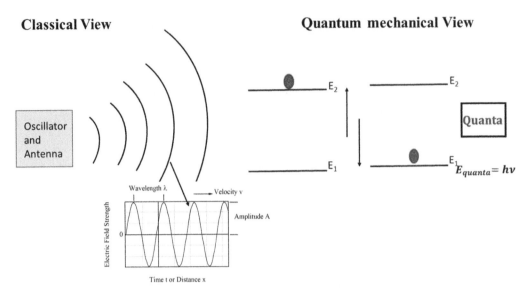

FIGURE 1.6   A schematic showing the differences in the classical and quantum mechanical view of radiation. According to classical electromagnetism, an oscillating charged particle, such as electron oscillating with some frequency, emits an electromagnetic wave of the same frequency continuously, moving with the speed of light in a vacuum, whereas according to Planck, an oscillating charged particle emits radiation discontinuously. It emits radiation only when it transitions from a higher energy state to a lower energy state. The frequency of the radiation it emits depends on the energy difference between the higher and lower energy levels.

EXAMPLE 1.1

According to displacement law, the wavelength of maximum thermal energy from a body at temperature "T" is mathematically described as, $\lambda_{max} T = 2.898 \times 10^{-3}$ m.K. For a human body at a temperature of about 70°F, the wavelength of the thermal radiation emitted:

$$\lambda_{max} = \frac{2.898 \times 10^{-3} \text{ m.K}}{294 \text{ K}} \tag{1.18}$$

$$\lambda_{max} = 10.0 \times 10^{-6} \text{ m}$$

Thus, the wavelength of thermal radiation emitted by human body is about 10 microns.

PROBLEMS

1.1   Can a charged particle moving with a constant velocity produces an electromagnetic wave? Why or Why not?

1.2   Give a brief description of all the experiments discussed previously. Explain the observation of the experiments that cannot be explained by the theory of classical physics.

1.3   What is a black body? Give examples.

1.4   What is the black-body paradox? Explain. Describe the difficulties faced by classical electromagnetic theory in describing the properties of black-body radiation.

1.5   Why did Planck introduce discontinuities for the energy of the radiation? Explain Planck's hypothesis for explaining the black-body radiation paradox.

1.6   What would be the energy of radiation emitted by an oscillating atom when it transitions from higher energy state $E_3 \rightarrow E_1$? Assume the frequency of oscillation to be 10^{12} Hz.

1.4   THE PHOTOELECTRIC EFFECT

In 1887, Heinrich Hertz, a German physicist, discovered a new phenomenon that was later called the photoelectric effect. The observation of this phenomenon was significant in the development of quantum theory. According to his studies, when ultraviolet light is shined on two metal electrodes with voltage applied between them it causes the electrons to be ejected from the metal electrodes. The principle experimental observations of his studies could not be explained by the classical theory of electromagnetism. These observations were:

1. The kinetic energy carried by the electrons could not be increased by increasing the intensity of the ultraviolet light. The kinetic energy carried by the electrons depended on the frequency of the ultraviolet light. This observation was inexplicable by classical physics. According to classical physics, ultraviolet light is an electromagnetic wave, and by increasing the intensity of the electromagnetic wave, the energy carried by the wave increases, and thus, the wave must impart more energy to the electrons when it interacts with the electrons of the metal plate. Thus, the kinetic energy of the emitted electrons must increase when the intensity of the ultraviolet light is increased. Also, the kinetic energy of the electrons cannot depend on the frequency of the incident ultraviolet light.
2. The increase in the intensity of the ultraviolet light increases the electron current. This implies that the number of electrons emitted from the electrode plate increases as the intensity of the ultraviolet light (or any electromagnetic wave) is increased. This could not be explained by the classical physics.
3. There was no time lag between the emission of electrons and the arrival of ultraviolet light on the metal plate. This observation was inexplicable by classical physics.
4. If the frequency of the electromagnetic wave is below a certain value, then no electrons are emitted. According to classical physics, electromagnetic waves of any frequency can cause emission of electrons.

For several decades, physicists performed experiments with different types of electromagnetic waves, such as visible light, infrared, X-rays and gamma rays, and tried different materials such as liquid and gas. However, the same observations were obtained, and scientists struggled to find an explanation for the phenomenon. In 1905, Einstein, guided by these difficulties, advanced Planck's ideas for explaining this phenomenon. According to his hypothesis:

1. An electromagnetic wave comprised of particles called photons. The energy carried by each of these photons is given as:

$$E_{photon} = h\nu \tag{1.19}$$

where h = Planck's constant, and ν = the frequency of the electromagnetic wave. A ray of light, which is an electromagnetic wave, consists of a number of photons. Each photon carries the above energy, and the total energy of the ray of light that is incident on a surface per unit area per unit time can be given as:

$$E_{EM} = nh\nu \tag{1.20}$$

where n = the average number of photons incident on the surface per unit area per unit time.
2. When a ray of light that consists of a number of photons is incident on a metal plate, the photons penetrate the plate and impart their energy to the electrons of the metal plate.

The electrons that have acquired this energy have to overcome a potential barrier called as a work function ϕ and then are released from the metal plate. The kinetic energy of each electron that is released from the metal surface is:

$$E_{K.E} = h\nu - \phi \tag{1.21}$$

where $E_{K.E}$ = the maximum kinetic energy of the electron emitted from the metal plate.

Using the hypothesis proposed by Einstein, the inexplicable observations can now be explained. The photons having particle nature impart all their energy to the electrons, which then overcome the potential barrier and are released from the metal surface and causes a photocurrent to be observed. The electrons released from the surface of the metal plate are called photoelectrons, and the current produced by them is called photocurrent. By increasing the intensity of the electromagnetic wave, the number of photons carried by such a wave increases. Thus, a larger number of photons leads to more electrons gaining sufficient energy to be released from the metal plate. This explains the cause of an increase in photocurrent when the intensity of the electromagnetic wave is increased. Equation (1.19) clearly illustrates the relationship between the frequency of an electromagnetic wave and the energy carried by a photon. Thus, by increasing the intensity of an electromagnetic wave, the kinetic energy carried by the electrons released from the metal surface cannot be increased.

Conceptual Question 2: Consider Equation (1.20) as it describes the energy carried by an electromagnetic wave. When you increase the intensity of the same electromagnetic wave, which parameter in the equation changes? Explain why.

The explanation of the photoelectric effect led to understanding that light could be made up of small particles called photons, and each photon carries a quantum of energy, as proposed by Planck. However, certain behaviors of light such as diffraction, dispersion and interference can only be explained by treating light as an electromagnetic wave. When particle-like behavior is considered for explaining these phenomenon, qualitative and quantitative difficulties are encountered. Thus, this understanding of the photoelectric effect led to advancement in the understanding of light, which never existed before. In 1921, Einstein received the Nobel Prize for explaining the photoelectric effect. However, it also shattered our view of light as "simply as an electromagnetic wave." In simple words, light is something more complicated than just a wave or a particle. Some of its phenomenon can be explained using the wave nature of light, while others can be explained using the particle nature of light. But why? Why does light behave so differently in different places? This is one of most difficult question that is still insufficiently understood (Figure 1.7 and Table 1.1).

Conceptual Question 3: Using Einstein's hypothesis, explain the third inexplicable observation of the photoelectric effect.

If the body is charged to a positive potential Π and surrounded by zero potential conductors, and if Π is just able to prevent the loss of electricity by the body, we must have

$$\Pi\varepsilon = \frac{R}{N}\beta v - P,$$

where ε is the electrical mass of the electron, or

$$\Pi E = R\beta v - P',$$

where E is the charge of a gram equivalent of a single-valued ion and P' is the potential of that amount of negative electricity with respect to the body.†

If we put $E = 9 \cdot 6 \times 10^3$, $\Pi \times 10^{-8}$ is the potential in Volts which the body assumes when it is irradiated in a vacuum.

To see now whether the relation derived here agrees, as to order of magnitude, with experiments, we put $P' = 0$, $v = 1 \cdot 03 \times 10^{15}$ (corresponding to the ultraviolet limit of the solar spectrum) and $\beta = 4 \cdot 866 \times 10^{-11}$. We obtain $\Pi \times 10^7 = 4 \cdot 3$ Volt, a result which agrees, as to order of magnitude, with Mr. Lenard's results.[3]

If the formula derived here is correct, Π must be, if drawn in Cartesian coordinates as a function of the frequency of the incident light, a straight line, the slope of which is independent of the nature of the substance studied.

As far as I can see, our ideas are not in contradiction to the properties of the photoelectric action observed by Mr. Lenard. If every energy quantum of the incident light transfers its energy to electrons independently of all other quanta, the velocity distribution of the electrons, that is, the quality of the resulting cathode radiation, will be independent of the intensity of the incident light; on the other hand, ceteris paribus, the number of

† If one assumes that it takes a certain amount of work to free a single electron by light from a neutral molecule, one has no need to change this relation; one only must consider P' to be the sum of two terms.

FIGURE 1.7   A clipping from Einstein's original paper on photoelectric effect. The equations shown in the text are now known as Equation (1.20) in the simpler version. The text that follows is derived from Einstein's original paper on photoelectric effect. The text provides an insight into Einstein's view of the light during those puzzling times. This paper was copied from the website The Collected Papers of Albert Einstein, http://einsteinpapers. press.princeton.edu/.

Another clipping from the same paper is given below. The last four lines of the clipping clearly illustrate new viewpoint of a ray of light in space.

The wave theory of light which operates with continuous functions in space has been excellently justified for the representation of purely optical phenomena and it is unlikely ever to be replaced by another theory. One should, however, bear in mind that optical observations refer to time averages and not to instantaneous values and notwithstanding the complete experimental verification of the theory of diffraction, reflexion, refraction, dispersion, and SO on, it is quite conceivable that a theory ai'light involving the use of continuous functions in space will lead to contradictions with experience, if it is applied to the phenomena of the creation and conversion of light. In fact, it seems to me that the observations on "black-body radiation", photoluminescence, the production of cathode rays by ultraviolet light and other phenomena involving the emission or conversion of light can be better understood on the assumption that the energy of light is distributed discontinuously in space. According to the assumption considered here, when a light ray starting from a point is propagated, the energy is not continuously distributed over an ever increasing volume, but it consists of a finite number of energy quanta, localised in space, which move without being divided and which can be absorbed or emitted only as a whole.

EXAMPLE 1.2

A light with a wavelength of about 10^{-7} m strikes a potassium metal plate. Determine the velocity of the photoelectrons released from the plate.

From Equation (1.20), we obtain:

$$\frac{1}{2}m_e v_e^2 = \frac{hc}{\lambda} - \phi$$

$$v_e = \sqrt{\frac{2}{m_e}\left(\frac{hc}{\lambda} - \phi\right)} \tag{1.22}$$

$$= \sqrt{\frac{2}{0.91 \times 10^{-30}\,\text{kg}}\left(\frac{6.63 \times 3 \times 10^{-26}\,\text{m.J}}{10^{-7}\,\text{m}} - 2.2 \times 1.6 \times 10^{-19}\,\text{J}\right)} \tag{1.23}$$

$$\Rightarrow v_e = 19 \times 10^5\,\text{m/s}$$

PROBLEMS

1.7 What wavelength of light is necessary to produce electrons, also called photoelectrons, from a sodium metal plate moving with a speed of about $0.001c$ ($c =$ the velocity of light)?

1.8 How many photons are released per second from a laser whose wavelength is 680 nm and has an operating power of about 10 mW?

1.9 Light of 350 nm strikes a metal plate, and photoelectrons are produced, moving as fast as 2.0×10^5 m/s. Determine the work function of the metal plate.

1.10 Using Table 1.1, determine the wavelength of incident light for which no electrons will be released from the zinc metal plate.

TABLE 1.1

Values of Work Functions of Different Metals

Metal	Work Function (eV)
Tungsten	4.5
Chromium	4.4
Zinc	4.3
Magnesium	3.7
Sodium	2.3
Potassium	2.2
Cesium	1.9

1.5 THE COMPTON EFFECT

The Compton effect was observed in 1923 by Arthur Holly Compton. He demonstrated another experimental observation toward the validation of the particle nature of light. The experiment consisted of directing radiation of high frequency such as X-rays on free electrons. The interaction between the X-rays and the free electrons lead to deflection of the X-rays with reduced frequency.

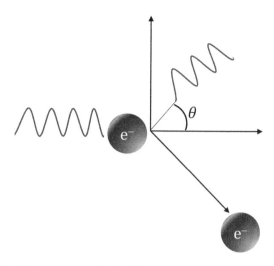

FIGURE 1.8 A schematic of electron scattering by X-rays.

Also, the interaction caused the electrons to recoil so that the electron seemed to be deflected in the same manner as if they had undergone collision with another particle. This phenomenon can be simply explained using the particle nature of light. An X-ray is comprised of X-ray photons, and each photon has energy, $E_{photon} = h\nu$ and they also carry momentum as shown:

$$p = \frac{h\nu}{c} = \frac{h}{\lambda} \tag{1.24}$$

The previous relationship is called De Broglie's momentum relationship for the matter waves associated with a material particle which will be discussed in details in the next chapter. Here p is the momentum carried by X-rays photon, and is the wavelength of the X-rays. Thus, the interaction between X-rays and free electrons can be viewed as a collision between X-ray photons (particles) and electrons, just in the same manner as an elastic collision between two billiard balls. The application of mechanical laws of conservation of momentum and energy lead to the establishment of the following relationship:

$$\frac{1}{\nu'} - \frac{1}{\nu} = \frac{h}{m_e c^2}(1 - \cos\theta) \tag{1.25}$$

where ν' = the frequency of the deflected photon after collision, ν = the frequency of the incident photon, θ = the angle of deflection, m_e = the mass of an electron, and h = Planck's constant. According to the previous equation, the loss in momentum of the incident X-ray photon during collision results in the emergence of a new X-ray photon of frequency ν' that is deflected at an angle θ with an incident photon. It also causes the stationary electron to gain momentum and rebound. The Compton effect further established the foundation of the particle nature of all types of radiation (Figure 1.8).

Conceptual Question 4: Explain, why did X-rays behave as a particle in the experiment of Compton scattering?

EXAMPLE 1.3

An X-ray photon of wavelength 0.0300 nm strikes a free, stationary electron, and a scattered photon is deflected at 90° from the initial position. Determine the momentum of the incident and scattered photon.

For the incident photon:

$$p_i = \frac{h}{\lambda} = \frac{6.63 \times 10^{-34} \, \text{J.s}}{0.0300 \times 10^{-9} \, \text{m}} = 2.21 \times 10^{-23} \text{kg.m/s} \tag{1.26}$$

The momentum of the deflected photon can be obtained by using Equation (1.25)

$$\lambda' - \lambda = \frac{h}{m_e c}(1 - \cos \theta) \tag{1.27}$$

$$\lambda' = \lambda + \frac{h}{m_e c}(1 - \cos 90) \tag{1.28}$$

$$= 3.0 \times 10^{-11} + \frac{6.63 \times 10^{-34} \, \text{J.s}}{9.1 \times 3 \times (10^{-31+8}) \, \text{kg.m/s}} \rightarrow 3.24 \times 10^{-11}$$

The momentum of the scattered photon:

$$p_{sc} = \frac{h}{\lambda} = \frac{6.63 \times 10^{-34} \, \text{J.s}}{0.0324 \times 10^{-9} \, \text{m}} = 2.04 \times 10^{-23} \text{ kg.m/s} \tag{1.29}$$

PROBLEMS

1.11   Calculate the energy of a photon of an electromagnetic radiation of frequency 10^{17} Hz. Also, determine the momentum carried by the photon.

1.12   Calculate the change in frequency of radiation that is incident upon an electron and deflected at angle of 30°. If the frequency of the incident photon is 10^{21} Hz, what would be the frequency of the deflected photon?

1.13   What would be the momentum carried by X-ray photons if the frequency of the X-rays is 2×10^{18} Hz.

1.6 TUTORIAL

1.6.1 PHOTOELECTRIC EFFECT

Purpose

To help students understand the particle behavior of an electromagnetic wave and show that the photoelectric effect cannot be explained using classical wave model. Use simulation to gain better understanding.

Concepts

- Classically, energy carried by an electromagnetic depends on its intensity. When an electromagnetic wave strikes a metal plate, it imparts energy to the electrons of the plate. The more intense the wave, the more energy is imparted to the electrons.
- By increasing the intensity of the electromagnetic wave, the kinetic energy of the emitted electrons must increase.
- Experimentally, it was observed that the kinetic energy of the electrons emitted from the plate depends on the frequency of the incident electromagnetic wave, not on its intensity.
- Classically, an electromagnetic wave of any frequency can cause the emission of electrons.
- Experimentally, it was observed that electromagnetic waves of certain frequencies only cause the emission of electrons.

Consider the Figure 1.9 that follows:

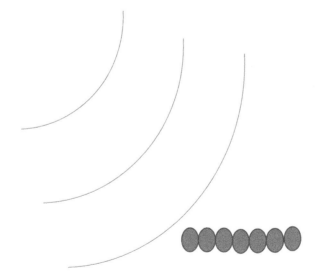

FIGURE 1.9 A schematic showing that an electromagnetic wave strikes a group of electrons.

Question 1 The intensity of light or an electromagnetic wave is the power transferred per unit area. It is proportional to the square of the amplitude of the electric field of an electromagnetic wave and is given as:

$$I \, \alpha \, |E|^2 \tag{1.23}$$

Using the previous equation, explain why waves of different frequencies but the same amplitude transfers same amount of power per unit area to a metal plate. Draw diagrams similar like Figure 1.9 to show that waves of different frequencies but same amplitude that strikes the electrons imparts the same amount of power or energy to the electrons. In your diagrams use oscillatory waves rather spherical waves.

_____

_____

_____

_____

_____

Question 2   Using the previous classical model of an electromagnetic wave, explain what would happen to the electrons energy after electromagnetic waves strike them continuously. Will the electrons gain or lose energy from the waves? Consider that the electrons are static previously.

_____

_____

_____

_____

Question 3   Consider that the waves of different frequencies but same intensity are shined on the electrons. What would be the difference in the energies of the electrons? Explain.

_____

_____

_____

_____

_____

FIGURE 1.10   This figure shows collision between two massive particles.

Consider Figure 1.10, and assume that the first particle's mass is m_1, $m_1 \ll m_2$, and initially moving with velocity v. The second particle is stationary. Assume that the collision is purely inelastic (the particles stick together).

Question 4   After the collision between the particles (Figure 1.10), what would be their final velocities? Write the expression for the final kinetic energy of the particles by considering loss of some energy during the collision. Let E_f = the final kinetic energy of the particles, E_i = the initial kinetic energy, ϕ = the loss of energy ($\Delta E = -\phi$), and $\Delta E = E_f - E_i$.

_____

_____

_____

_____

_____

Question 5   Apply the previous particle model to an electromagnetic wave by assuming that electromagnetic wave comprises of particles (in the case of light, particles are called "photons"), and each particle carries energy

$$E_{particle} = h\nu$$

Now, using previous equation and the equation for the kinetic energy that you have obtained in **Question 4**, derive Einstein's kinetic energy relationship (Equation 1.21). What are the new assumptions? What is the difference between these assumptions and the classical particle model?

_____

_____

_____

_____

_____

Simulation

Use the following simulation for enhancing your understanding of the photoelectric effect.
https://phet.colorado.edu/en/simulation/photoelectric
On the right of the screen, choose sodium as metal plate.

Question 5 Starting from the infrared (IR), determine the wavelength of the electromagnetic wave at which electrons are released from the metal plate. Determine the wavelength at which electrons are not released from the metal plate.

Increase the intensity (keep the wavelength constant) and record your data (keep the voltage of battery to 0 V).

Intensity	Current (A)
10%	0
20%	0.05

Do you observe an increase in current because of increase in the intensity? Explain why.

_____

_____

_____

_____

_____

Do you observe an increase in energy of the electrons as the intensity of the electromagnetic wave is increased? (For this, look up for the electron energy vs frequency graph on the right side of the screen.)

Now, keep the intensity constant, and increase the wavelength and record your data for wavelength vs current.

Wavelength (nm)	Current (A)
300	0
520	0.1

What do you observe from the above data? Explain your observations by using Equation (1.20).

_____

_____

_____

_____

_____

What do you observe when you increase the voltage of the battery by a small amount? Explain your observations.

_____

_____

_____

_____

_____

2 An Introductory History of Quantum Mechanics-II

2.1 DE BROGLIE MATTER WAVES

In the last chapter the experiments that led to key understandings concerning the characteristics of radiation were elucidated to explain that it has a discrete energy or it behaves as a particle. The black body radiation and photoelectric effect experiments demonstrated that electromagnetic radiation such as light behaves as a particle. Long before Planck's discovery, it was established by experiments on interference, diffraction and many other phenomena in optics that radiation is a wave. This new discovery was contrary to this well-established belief about radiation. Does radiation have a dual nature? In some experiments radiation behaved as a wave and in some experiments, it behaved as a particle. Why? What are the underlying causes for such behavior? If radiation can behave as a particle, what about the behavior of material particles? These questions were greatly puzzling to scientists during the beginning of the nineteenth century.

In 1924 Louis De Broglie tried to answer these questions and proposed a hypothesis which would form the basis of his Ph.D. thesis. De Broglie suggested that this dual nature of radiation may exist for matter particles as well. That is, material particles would also exhibit wave properties. At that time, the wave nature of material particles was not known. His idea was very original. Why was such wave aspect of material particles never revealed in any experiments before the advent of quantum theory? De Broglie suggested this could have been because the wavelength of material particles was so small that more sensitive devices than were available at the time would have been required to detect their existence. He developed a mathematical theory of matter waves that explained the behavior of such waves and the phenomena exhibited by them.

De Broglie began with an examination of the behavior of a particle moving freely in space. Such a particle whose mass is "m_0" would move in a straight line with a constant speed "v." The relativistic expressions of such a particle's energy and momentum are:

$$E = \frac{m_0 c^2}{\sqrt{1 - \dfrac{v^2}{c^2}}}; \quad p = \frac{m_0 v}{\sqrt{1 - \dfrac{v^2}{c^2}}} \tag{2.1}$$

where p is the momentum of the particle, c is the speed of light in a vacuum, m_0 is the particle mass, and v is the constant speed of the particle.

According to De Broglie, a corresponding plane wave would be associated with such a particle. The energy and momentum of such a particle also must be related to the energy relationship obtained by Einstein for light wave photons. Therefore, the energy of the particle E is:

$$E = \frac{m_0 c^2}{\sqrt{1 - \dfrac{v^2}{c^2}}} = h\nu \tag{2.2}$$

By assuming that the energy and momentum of the particle are related to the frequency and wavelength of the wave by the same relations that hold for photons and waves of light, the velocity (V) of such a wave can be derived:

$$V = \nu\lambda \tag{2.3}$$

De Broglie postulated that the momentum of the particle and frequency and velocity of the wave associated with a particle can be

$$E \approx pV \rightarrow p = \frac{E}{V} \tag{2.4}$$

Since $E = h\nu$.

Therefore,

$$p = \frac{h\nu}{V} \tag{2.5}$$

and using Equation (2.3),

$$p = \frac{h}{\lambda} \tag{2.6}$$

The above relationship is called De-Broglie's momentum-wave relationship. It relates the momentum of a particle moving freely in space with the wavelength of the wave associated with the particle.

From Equation (2.1), considering the relativistic equations, the velocity of the wave associated with the particle is,

$$\frac{p}{E} = \frac{v}{c^2}; \quad \text{and} \quad \frac{p}{E} = \frac{(h/\lambda)}{h\nu} = \frac{1}{V} \tag{2.7}$$

and thus,

$$\frac{1}{V} = \frac{v}{c^2} \rightarrow V = \frac{c^2}{v} \tag{2.8}$$

Here, 'V' is the velocity of the De-Broglie wave associated with the particle that is moving freely in space obtained from the relativistic equations. The wave crests move faster than the particle and in fact can move faster than the velocity of light in the medium (including in a vacuum). Is this feature of De Broglie's wave in conflict with the theory of relativity? The theory of relativity prohibits velocities greater than 'c' (c is the speed of light in a vacuum) as long as these velocities refer to the transfer of energy. The De Broglie's waves transport no energy, and the energy of the particle remains constant. Thus, there is no conflict with the theory of relativity. Also, if the particle becomes stationary, that is $v = 0$, then according to Equation (2.8), the velocity of the wave becomes infinite. For such a situation, the sinusoidal wave turns into a standing wave around the particle. However, a wave moving faster than the particle remains contradictory. We will show later in Chapter 4, that the theory of quantum mechanics developed afterwards resolves all of these contradictions (Figure 2.1).

The above derivations reveal early concepts in the development of quantum mechanics. By deriving the relationship between the momentum of a particle and the wavelength of the wave associated with a particle, De Broglie laid the foundations of Wave Mechanics. With the advancements in the mathematical theory of wave mechanics, the above theory was advanced to include probability waves rather than electromagnetic waves associated with a particle and wave packets rather a planar sinusoidal waves associated with a particle. In the section below the derivation of the De Broglie

To start with, was the following striking remark : in relativity theory, the frequency of a plane monochromatic wave is transformed as $\nu = \nu_0/\sqrt{1 - \beta^2}$ whereas a clock's frequency is transformed according to a different formula : $\nu = \nu_0\sqrt{1 - \beta^2}$ ($\beta = v/c$). I then noticed that the 4-vector defined by the phase gradient of the plane monochromatic wave could be linked to the energy-momentum 4-vector of a particle by introducing h, in accordance with Planck's ideas, and by writing :

$$W = h\nu \qquad p = h/\lambda \qquad (1)$$

where W is the energy at frequency ν, p the momentum, and λ the wavelength. I was thus led to represent the particle as constantly localized at a point of the plane monochromatic wave of energy W, momentum p, and moving along one of the rectilinear rays of the wave.

However, and this is never recalled in the usual treatises on Wave Mechanics, I also noticed that if the particle is considered as containing a rest energy $M_0 c^2 = h\nu_0$ it was natural to compare it to a small clock of frequency ν_0 so that when moving with velocity $v = \beta c$, its frequency different from that of the wave, is $\nu = \nu_0\sqrt{1 - \beta^2}$. I had then easily shown that while moving in the wave, the particle had an internal vibration which was constantly in phase with that of the wave.

FIGURE 2.1   A reproduction of a portion of De-Broglie's original paper on wave mechanics.

relationship using a wave packet will be discussed without considering relativistic effects. Figure 2.2 below depicts a wave associated with a particle.

Conceptual Question 1: If matter particles have waves associated with them then why weren't these waves observed before Planck's findings? Explain using De-Broglie's relation.

Wave Packet: A particle is well localized in space, and a wave spreads through space. Thus, associating a wave with a particle implies that the wave is localized because particle must be located

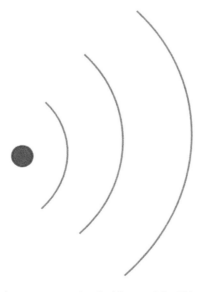

FIGURE 2.2   A schematic depicting waves associated with a particle. This schematic is mainly to aid in visualization of the waves associated with a particle.

somewhere in space. It is conceptually quite contradictory to associate a wave with a particle. This logical difficulty can only be removed by assuming that the wave associated with a particle can only be represented as a wave packet. A wave packet is comprised of a group of waves of slightly different wavelengths, with corresponding phases and amplitudes that interfere constructively over only a small region of space. Outside of that region, they produce an amplitude that rapidly diminishes to zero as a result of destructive interference. Figure 2.3 depicts such a wave packet.

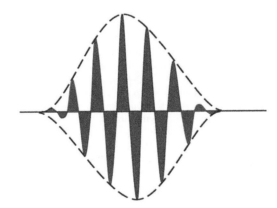

FIGURE 2.3   A wavepacket comprised of superimposed waves. This image was produced by Haocheng Yin.

The regions of constructive interference coincide with the location of the particle, while conversely the particle does not exist in regions of destructive interference. Thus, a wave packet can be used to describe the behavior of a single particle. The motion of the wave packet is caused by the change of phase of all the different wavelengths. The change of position of the wave packet is caused by the change of conditions for destructive and constructive interference which leads to a change in the position of a wave packet. The velocity with which individual waves of a wave packet moves is called phase velocity.

De Broglie's relationship will be obtained by associating a wave packet with a particle. The velocity with which a group of waves moves in space is called the "group velocity" of the wave packet. It is same as the particle's velocity such that:

$$v_g = \frac{d\omega}{dk} = \frac{p}{m} \tag{2.9}$$

where p is the particle's momentum. Since,

$$E = h\nu = \hbar\omega \rightarrow \frac{d\omega}{dk} = \frac{1}{\hbar}\frac{dE}{dk} = v_g \tag{2.10}$$

Classically, the energy of the particle is:

$$E = \frac{p^2}{2m} \rightarrow \frac{dE}{dk} = \frac{p}{m}\frac{dp}{dk} \tag{2.11}$$

Using all of the above equations, the following can be derived:

$$\frac{p}{m} = \frac{1}{\hbar}\frac{p}{m}\frac{dp}{dk} \tag{2.12}$$

Thus, by integrating the above equation, the following can be derived:

$$p = \hbar k = \frac{h}{\lambda} \tag{2.13}$$

This equation is de Broglie's relation. The constant of integration is set to zero as a method of simply describing a particle's motion by a wave packet. The wavelength is the average wavelength of the wave packet. The classical expression for the energy of a free particle is:

$$E = \frac{p^2}{2m} = \frac{(mv_g)^2}{2m} = \frac{1}{2}mv_g^2 \tag{2.14}$$

Therefore, the group velocity of the wave packet is the velocity with which the particle moves in free space.

Conceptual Question 2: A wave packet comprises of the superposed waves of different wavelengths. What is the wavelength of a wave packet?

Conceptual Question 3: A particle is localized in space whereas wave spreads over space. Thus, a wave associated with a particle is contradictory. Explain, how this problem can be resolved by using a wave packet?

PROBLEMS

2.1 Determine the wavelength of a wave associated with an electron that is moving at a velocity of 8×10^6 m/s, and of a wave associated with a ball moving at a velocity of 20×10^3 m/s? Assume the mass of the ball to be 2 kg. Compare these wavelengths and explain why the wave associated with the ball cannot be observed.

2.2 A photon has the same momentum as an electron moving with a speed of 40×10^5 m/s. Find the wavelength of the photon. Compare the kinetic energies of the photon and the electron.

2.3 Consider that the width of a wavepacket of an electron (Δx) spreads in time. The spread in time is specified by the following relation:

$$\Delta x = \Delta x_0 \sqrt{1 + \frac{\hbar^2 t^2}{m^2 (\Delta x_0)^4}}$$

Determine the time required for the wavepacket to expand to twice its original width?

2.2 ELECTRON DIFFRACTION

De Broglie's hypothesis that there are waves associated with material particles was confirmed experimentally by Clinton Davisson and Lester Germer in 1927. They demonstrated that material particles such as electrons exhibit diffraction and interference phenomenon. This experimental validation of wave-particle duality was an important step towards the development of wave mechanics.

Classically, when a wave passes through a slit or obstacle whose width is comparable to the wavelength of the wave. The wave bends around the corners of the slit. This bending of light is called diffraction. The bending of light produces interference of the waves. This interference can be observed as the maxima and minima of intensity of the wave impinging on a reflective screen. All kind of waves such as electromagnetic waves, sound waves and water waves exhibit diffraction phenomenon. Figure 2.4 below describes this diffraction phenomenon is more detail.

In the experiment of Davisson and Germer, an adjustable energy electron beam impinged upon a nickel crystal surface, and the current of the scattered electrons was measured at a particular scattering angle. The planar surface of the nickel crystal is comprised of scattering surfaces at regular intervals in a similar manner to a diffraction grating. The waves that scatter from one planar surface can constructively or destructively interfere with waves from secondary crystal planes "deeper" in the crystal. The experimental set up of their experiment is depicted in the figure below. The waves associated with the electrons exhibited diffraction and produced interference pattern in a directly analogous manner to light photons.

The periodic structure of the crystal serves as a three-dimensional diffraction grating, and the angles of maximum reflection are Braggs's condition for constructive interference, which is known as Bragg's law:

$$n\lambda = d \sin(\theta) \tag{2.15}$$

where λ is the wavelength, d is the spacing between the crystal planes, and θ is the scattering angle.

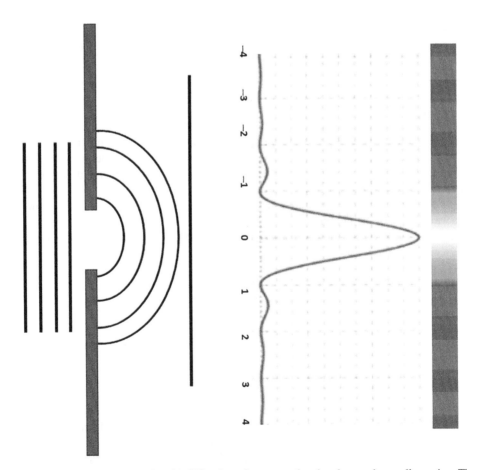

FIGURE 2.4 A schematic illustrating the diffraction of waves as they bend around a small opening. The corresponding maxima and minima demonstrate the spreading of the intensity of the wave.

The waves associated with the electrons undergo diffraction and produce interference patterns. The intensity of the waves will be maximized at a certain scattering angle. In the above equation for $n = 1$, $d = 2.15$ A (Angstrom) spacing between the planes of the nickel atoms, $\theta = 50°$, the wavelength of the wave undergoing diffraction is 0.165 nm. If the kinetic energy of the electron beam incident on the nickel crystal is 54 eV (or 54 volts), then applying De- Broglie relation, the wavelength of the electron waves can be derived as:

$$\lambda = \frac{h}{\sqrt{2mE}} \tag{2.16}$$

This measured wavelength was 0.167 nm for nickel, which is approximately equal to the wavelength obtained from Bragg's law (Equation 2.15). Thus, Davisson and Germer's experiment confirmed the existence of matter waves associated with electrons by demonstrating that the angles at which electrons are reflected matched the patterns that are evident in X-ray diffraction. Since X-rays are electromagnetic waves that exhibit diffraction and interference effects, it was a very important step toward the development of quantum mechanics. Both Davisson and Germer received the Nobel prize in 1937 for this discovery.

Conceptual Question 4: How did Davisson and Germer confirmed their findings that electrons have waves associated with them? Explain.

PROBLEMS

2.4 Show that De Broglie's formulae for a beam of electrons can be approximately written as,

$$\lambda\,(Angstrom) = \sqrt{150/V}$$

where V is the voltage being applied to accelerate the electrons.

2.5 Determine the wavelength of a beam of electrons incident on a nickel crystal that is scattered at an angle of $\theta = 28°$ when the potential applied to accelerate the electrons is 106 V.

2.3   DOUBLE-SLIT EXPERIMENT

Thomas Young's double-slit experiment during the eighteenth century confirmed the wave theory of light. It showed that light is comprised of a wave that is split into two when passing through a double-slit and that these two waves interfere constructively and destructively to produce interference phenomenon. Thus, the intensity of light that has undergone interference splits into maxima and minima of the intensity when observed on a screen. Figure 2.5 below illustrates the interference phenomenon of light.

If there are matter waves associated with electrons, then they must exhibit interference phenomenon when electrons pass through a double-slit. Therefore, performing Young's experiment for electrons must demonstrate interference phenomenon associated with the matter waves of electrons.

The existence of matter waves of electrons was further confirmed by Clauss Jonsson in 1961. He was the first experimentalist to demonstrate that electrons exhibit interference phenomenon when passing through a double-slit. In 1974, Pier Giorgio, GianFranco Missiroli and Giulio Puzzi also demonstrated that electrons passing through a single slit exhibit diffraction. They also demonstrated that a single electron interferes with itself when the intensity of the electron beam reduced so that a single electron passes through each slit one at a time during the double-slit experiment. In 1989, a team led by Akira Tonomura at Hitachi performed a double slit experiment. For this experiment, each single

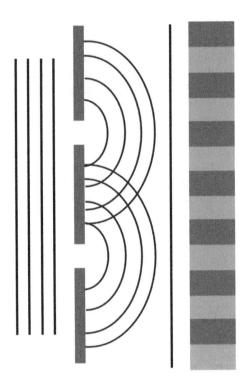

FIGURE 2.5   The interference of waves as they enter through a double slit.

electron passed through a single slit one at a time and arrived at the screen of a detector as a single particle as a "dot." The location of each dot is then recorded for each electron transit. Since each electron passes through the slit singularly, there is no way one electron can interfere with the other electron. The dots appearing on the screen means that electrons behave as particles when they are detected at the screen. However, each electron arrives at the screen at a different location. After many more electrons strike the screen, the interference pattern on the screen emerges, and maxima and minima are observed. The regions on the screen where the highest concentration of electrons impinge is called the maxima, and the regions where the least numbers of electrons arrive is called the minima. The existence of these clear distinct regions demonstrate that each electron has interfered with itself to produce such an interference pattern. This experiment is remarkable in a sense that it helps to illustrate the quantum mechanical features so closely and shows the distinctive nature of quantum mechanics in comparison to classical physics. The figure below illustrates such interference phenomenon for electrons (Figures 2.6 and 2.7).

The figure above clearly shows the interference pattern, with clear distinct regions of maxima and minima. Initially, with only a few electrons impinging on the screen, it is difficult to identify the interference pattern. As the number of electrons is increased, the interference pattern emerges more clearly. There are several questions that can be asked regarding the interference pattern produced by electrons. If only one electron passes through the slit at a time, then why do they arrive at different locations on the screen? Since, according to classical physics, if each electron passes through a single slit and arrives at the screen as a dot or particle then we must observe only two locations on the screen where electrons are impinging. We do not observe these two spots, but rather a spread-out distribution of electrons, because each electron interferes with itself. This interference causes the spreading of electron distribution. What is the mechanism for an electron interfering with itself? If an electron interferes with itself just like a wave, then why does it strike the screen as a dot (particle)? What happens to its wave nature at that point?

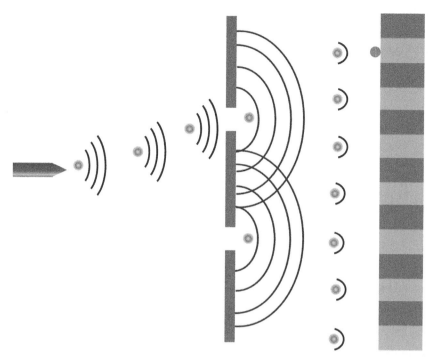

FIGURE 2.6 The above schematic illustrates the interference patterns of electron waves. The regions of maxima (light color) are the regions where electrons are most likely to arrive (strike the screen). The electrons arrives in these regions and form interference pattern.

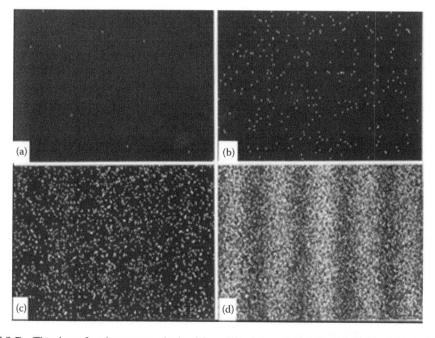

FIGURE 2.7 The above four images are obtained from Hitachi's website http://www.hitachi.com/rd/portal-/highlight/quantum/doubleslit/index.html with permission from Hitachi. (These images are from a journal paper, A. Tonomura et al. *American Journal of Physics*, Vol. 57, No. 117, 1989.) These images demonstrate the inference of the electron waves. The interference pattern of the electrons gradually builds up (a)–(d).

To understand the double-slit experiment and the interference of waves in a little more detail, it is useful to review the conclusions of classical physics.

Consider Figure 2.5, for a wave entering slit 1, the mathematical function that describes such a wave is:

$$f_1(r, t) = A_1 e^{i(k_1 r - \omega t + \phi_1)} \tag{2.17}$$

where A_1 is the amplitude of the wave and ω is the angular frequency, k_1 is the wave vector. Similarly, the wave entering through slit 2 can be described as:

$$f_2(r, t) = A_2 e^{i(k_2 r - \omega t + \phi_2)} \tag{2.18}$$

The corresponding intensities of the two waves are the absolute squares of the above functions, and thus:

$$I_1 = |A_1|^2, \quad I_2 = |A_2|^2 \tag{2.19}$$

After entering slit 1 and slit 2, these two waves superpose on each other:

$$f(r, t) = A_1 e^{i(k_1 r - \omega t + \phi_1)} + A_2 e^{i(k_2 r - \omega t + \phi_2)} \tag{2.20}$$

and as a result their intensities sum as follows:

$$I_{12} = |f(r, t)|^2 = |A_1|^2 + |A_2|^2 + 2 A_1 A_2 \cos\theta \tag{2.21}$$

where:

$$\theta = (k_1 - k_2) \cdot r + (\phi_1 - \phi_2) \tag{2.22}$$

Such that:

$$I_{12} \neq I_1 + I_2 \tag{2.23}$$

The total intensity reaches maxima of constructive interference, when:

$$\theta = 0, \pm 2\pi, \pm 4\pi, \ldots\ldots \tag{2.24}$$

and reaches minima of destructive interference, when:

$$\theta = \pm \pi, \pm 3\pi, \pm 5\pi, \ldots\ldots \tag{2.25}$$

The pattern of maxima and minima is produced on a screen due to the third term in Equation (2.21) which is called the "interference term." Interference vanishes if this term is removed from the above equation and two intense spots would be observed as shown in the Figure 2.8.

With this understanding let us turn to an understanding of quantum mechanical interference. Consider Figure 2.6, in which an electron gun emits electrons such that each electron either enters through slit 1 or slit 2 at a given time. The wave function of an electron that has matter wave associated with it is a complex function and very similar to the function in Equation (2.17). The electron wave function can be described as:

$$\psi_1(r, t) = B_1 e^{i(k_1 r - \omega t + \phi_1)} \tag{2.26}$$

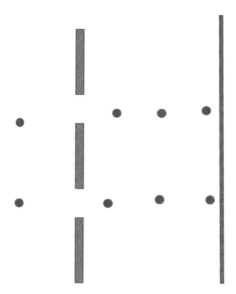

FIGURE 2.8 The schematic shows that if instead of waves, particles are made to pass through a double-slit then two intense spots will be observed on a screen.

The above equation describes the wave function of the electron entering through slit 1. Here B_1 is not the amplitude of the intensity of the matter wave of the electron, but rather is referred to as the "probability amplitude." The matter wave of a particle can also be referred to as a "probability wave." The absolute square of the wave function is termed the 'probability density' as listed below in Equation (2.27),

$$P_1(r) = |\psi_1(r, t)|^2 \tag{2.27}$$

$P_1(r)dr$ is the probability that an electron entering slit-1 can be found between r and $r + dr$. The wave function for the electron entering through slit 2 is:

$$\psi_2(r, t) = B_2 e^{i(k_2 r - \omega t + \phi_2)} \tag{2.28}$$

where

$$P_2(r) = |\psi_2(r, t)|^2 \tag{2.29}$$

is the corresponding probability density.

$P_2(r)dr$ is the probability that an electron entering slit-2 can be found between r and $r + dr$. The probability densities of the electrons that have entered through slit 1 or slit 2 are:

$$|\psi_1|^2 = |B_1|^2, \quad |\psi_2|^2 = |B_2|^2 \tag{2.30}$$

The probability distribution, which is also called the probability density of the electrons must be the sum of these probability densities. Thus:

$$|\psi_1|^2 + |\psi_2|^2 = P_{12} \tag{2.31}$$

However, if the probability densities add up in this way then there will be no observation of the interference of the probability distribution. In such a scenario, only two spots would be observed on the screen with no interference pattern. This is the same pattern that described above for particles.

Since the specific slit through which the electron has entered is not known, the wave function of the electron entering through the double slit will be the sum of the functions above. The superimposed wave function is:

$$\Psi(r) = \psi_1(r) + \psi_2(r) \tag{2.32}$$

and the probability densities of the electron add up in the same manner as waves as defined by the following relation:

$$|\Psi(r)|^2 = |\psi_1 + \psi_2|^2 = |B_1|^2 + |B_2|^2 + 2\,|B_1||B_2|\cos\theta \tag{2.33}$$

Equation (2.33) describes the probability distribution of the electron on the screen, and the third term is very similar to the interference term of Equation (2.21). This term causes uncertainty about where the electron will arrive on the screen. The probability distribution of the electrons has the interference pattern of maxima and minima as a result of this term. The electrons arrive on the screen as particles or dots, however their probability distribution on the screen behaves similarly to the intensity distribution pattern of waves.

Conceptual Question 5: *Explain the mathematical similarity between the intensities of the superposed waves of light and the probability densities of the superposed probability waves associated with a particle?*

Conceptual Question 6: *Is matter wave same as probability wave? Explain.*

PROBLEMS

2.6 Derive Equation (2.21) using Equation (2.20). Confirm that for monochromatic waves moving through the same medium, the spatial dependence in the phase factor (Equation 2.22) vanishes.

2.7 List brief descriptions for all the experiments discussed above together with their interpretations.

2.4   UNCERTAINTY PRINCIPLE

After this discussion of the wave properties of matter, it is appropriate to approach the most fundamental principle of quantum mechanics, the Uncertainty Principle. According to this principle, no matter how accurately the variables of a quantum mechanical system are determined, a fundamental uncertainty concern their actual magnitude will always remain. It is not possible to gain absolute deterministic information concerning a quantum system. This principle provides an absolute limitation on a determination of the value of a given quantum variable.

For example, consider the double slit experiment. In order to determine which slit the electron passes through, let us place a detector at slit 1 that shines a light or any other radiation on the electron to determine whether the electron passes through the slit (position of electron), as shown in Figure 2.9.

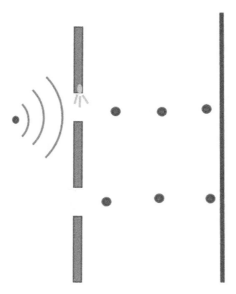

FIGURE 2.9 The above schematic illustrates that when a detector is placed on one of the slits, the interference pattern vanishes. Instead, two bright spots are observed.

Since light is comprised of photons, as light is shone on the electron that passes through the slit 1, the collision between the photon and the electron knocks the electron from its path. The photon from the light source becomes scattered which can be registered as a detection of an electron. The kick that an electron receives from the photon changes its momentum. This changed momentum cannot be exactly known and is uncertain. However, the position of the electron has become certain. The scattered photon provides the information about the location of the electron entering through slit 1. It is the fundamental rule of nature if the position of a quantum system has the least uncertainty then momentum has the most uncertainty. Since the momentum and the wavelength of the wave associated with the electron are related by De-Broglie's relation, the wavelength becomes negligible as momentum becomes uncertain. Hence, all the electrons whose positions are determined follow the straight path because there is no probability wave associated with them and they strike the screen at approximately the same spot. The interference of the electron waves cannot be observed and so instead two intense spots are observed on the screen as depicted in Figure 2.8. Mathematically, in Equation (2.32), the wavefunction will no longer be a superimposed wavefunction, but rather a single function, $\psi_1(x)$, since, we exactly know exact location of the electron.

Why does the certainty in position result in the momentum being uncertain? This is mainly due to a fundamental principle of nature. The principle that mathematically describes such uncertainty, is known as the Uncertainty Principle.

2.4.1 The Uncertainty Principle

According to the uncertainty principle, position and momentum variables of a quantum system cannot be simultaneously determined. When a measurement is performed to measure such variables then uncertainty in their values will be observed. The relation that describes the uncertainty between position and momentum of a quantum system is as follows:

$$\Delta x \, \Delta p_x \geq \frac{\hbar}{2} \tag{2.34}$$

where Δx is the uncertainty in the position of a particle (which is also called the standard deviation), and Δp is the uncertainty in the momentum of the particle. This way of describing position and momentum uncertainties is a mathematically informal way of describing the uncertainty principle. In Chapter 3 a mathematically appropriate derivation of the uncertainty principle will be demonstrated.

Let us now apply this principle to the above double-slit experiment. By placing the detector at slit 1, the exact location of a particle can become known since it has entered either slit 1 or slit 2 and is detected when it does. As a result of this, the uncertainty in the position becomes negligible ($\Delta x \approx 0$) From the Uncertainty Principle if the uncertainty in the position is negligible then the uncertainty in momentum becomes infinite (Equation 2.34). According to the De-Broglie relation, if momentum is uncertain (indeterminate) then the wavelength associated with the particle becomes negligible. It is like saying, "if we do not know the momentum of a particle, we cannot know its wavelength." In physics, if we cannot measure a physical quantity, then we cannot say whether that quantity even exists. Which means that there is no probability wave associated with the electron. As a result of this, no interference of the probability wave is observed. Below a quotation from Feynman Lectures is presented. According to Feynman Uncertainty Principle is of great significance in quantum mechanics. Without this principle, the quantum mechanics will just collapse.

> The uncertainty principle "protects" quantum mechanics. Heisenberg recognized that if it were possible to measure the momentum and the position simultaneously with a greater accuracy, the quantum mechanics would collapse, so he proposed it must be impossible.
>
> **Richard Feynman**
> *The Feynman Lectures on Physics*

The above uncertainty relation can be extended to the other two dimensions of the space, and can be described as:

$$\Delta y\,\Delta p_y \geq \frac{\hbar}{2} \tag{2.35}$$

$$\Delta z\,\Delta p_z \geq \frac{\hbar}{2} \tag{2.36}$$

In a similar manner, if the energy of a quantum system is measured, then the uncertainty in the values of energy (ΔE) and the time (Δt) at which this measurement occurs must conform to the following relation:

$$\Delta E\,\Delta t \geq \frac{\hbar}{2} \tag{2.37}$$

But time is not a dynamical variable, and thus the uncertainty in time refers to the time that the system takes to change. Consider a wave packet which is simply a short pulse that passes by a point "P." Since the wave packet consists of waves of different frequencies, the range of energy of the wave packet is ΔE. Since the wave packet spreads with time, the time it takes to pass through a point "P" would be uncertain by a time Δt. Thus, according to the uncertainty relation:

$$\Delta t \approx \frac{\hbar}{2\Delta E} \tag{2.38}$$

In another example, consider a two-level atom, the two levels are in a ground state and an excited state, and an electron is in the excited state. Due to perturbation, the electron transitions from the excited state to the ground state. Here ΔE represents the uncertainty in the exact value of energy of the electron when it is transitioning to the ground state and Δt represents the uncertainty in time, the instant of time at which the electron transitions to the ground state. These two variables are related by the uncertainty relation. More precisely the energy of the electron is known when it is transitioning to the ground state, however, the time when it will be in the ground state is known less precisely.

2.4.2 A SIMPLE PROOF OF THE UNCERTAINTY PRINCIPLE

Consider a wave packet that comprises of Δv, $\Delta \omega$, a range of frequencies as, Δk as a range of wave numbers and Δx as the spread or width. The classical relation between the width of the wavepacket and range of wave numbers is:

$$\Delta x\, \Delta k \geq \frac{1}{2} \tag{2.39}$$

This means that a narrow wave packet in space will exhibit a broad range of wave numbers and vice-versa If this wave packet represents a quantum particle. Then, according to De-broglie's relation:

$$\Delta k = \frac{\Delta p}{\hbar} \tag{2.40}$$

By substituting Equation (2.40) in Equation (2.39), the following is derived:

$$\Delta x\, \Delta p \geq \frac{\hbar}{2} \tag{2.41}$$

Applying the concept outlined above of the wave packet will facilitate a more detailed understanding of the double-slit experiment. In the double slit experiment, if the position of the particle become certain, which means that the width of the wave packet is almost negligible, then the range of momentum become meaningless. For such a case, a quantum particle cannot be described as a wave packet. It can only be described as a classical particle.

Similarly, the energy time uncertainty relation can be obtained by considering a wave packet of a range of frequencies $\Delta \omega$ that requires time Δt to pass through a certain point. The classical relation between this time and range of frequencies is:

$$\Delta \omega\, \Delta t \geq \frac{1}{2} \tag{2.42}$$

If such a wave packet is a quantum particle, then the range of energies and frequencies follows the relation:

$$\Delta E = \hbar\, \Delta \omega \tag{2.43}$$

By substituting Equation (2.43) in Equation (2.42), the following is obtained:

$$\Delta E\, \Delta t \geq \frac{\hbar}{2} \tag{2.44}$$

EXAMPLE 2.1

Consider a quantum particle that is represented as a wave packet. As the wave packet moves in space and time, its width spreads to (Δx). What will happen to its momentum?

The momentum (p) and width of the wave packet are related by the uncertainty relationship.

$$\Delta x \, \Delta p \geq \frac{\hbar}{2}$$

The average momentum of the wave packet remains constant with time. However, the range of momentum (Δp) continues to change with time. If the width of the wave packet become large, then the range of momentum becomes narrower and vice-versa. Although the wave packet spreads with time, the uncertainty relationship is always maintained.

2.5 MEASUREMENT

In classical physics, measurement is the only way to gain information about the properties of a system. For example, to know the length of a table a meter stick can be used. Using this stick, how small or large the table is with respect to other tables can be determined. So, this meter stick (a measuring device) helps us to gain knowledge about the table, which simply becomes our reality about the table. The error in the measurement of the length of the table is negligible and the length of the table is independent of the meter stick. Whether you use a meter stick or another means of measuring the dimensions of the table, the size of the dimensions will remain the same and thus our reality. What if the act of measurement of the length of the table using a meter stick, changes the length of the table? What if before measurement, a table has a different length? What would be the reality of the table then? This conundrum illustrates the unpredictability of quantum mechanics. For example, consider the double-slit experiment that was discussed in the previous section. According to that experiment, an electron has a probability wave associated with it. When the electron is passed through a double-slit, its probability wave interferes, which produces maxima and minima on a screen in a similar manner to the interference of classical waves.

Now when a detector is placed on one of the slits to determine which slit the electron passes through, the interaction between the electron and the detector destroys the probability wave of the electron. Without a probability wave, the electron arrives on the screen just like a classical particle, and thus no interference is observed. Thus, the act of observing the electron has changed its behavior. It behaves like a classical particle if we employ a detector to observe it. If we do not employ such a detector, the electron behaves like a quantum particle by producing interference of the probability waves on the screen.

In other words, by placing a detector on one of the slits, the interaction between the detector and the electron causes it to behave like a classical particle and to follow a deterministic path. Otherwise, where it arrives on the screen is indeterminate. It will arrive on any of those regions of maxima. So, what is the reality about the electron? Is it a classical particle that follows a deterministic path or a quantum particle that can arrive anywhere on the screen, whose path is not deterministic? Is it that the act of measurement changes our reality about the electron? You may be wondering if interaction between the electron and the detector is a problem then why can't we build a detector whose interaction with the electron is very negligible, so that it does not disturb the electron in any way but still detects which slit the electron passes through? The limitations on such a device are set by the Uncertainty Principle. According to which there is no way to build a detector that can detect an electron without disturbing it. The uncertainty in the measurement is governed by the Uncertainty Principle. Thus, the Uncertainty Principle sets the limits on determinism. So far no one has been successful in overcoming the limitations set by the uncertainty principle. Perhaps it can be said that the Uncertainty Principle describes the most fundamental principle of nature. This may be precisely the reason that quantum mechanics has been such a successful theory. It has been able to describe the properties of matter and light in a way that was considered impossible through classical physics.

Thus, the behavior of a quantum system is not independent of its environment. It behaves in a certain way because of the influence of the environment in which quantum system exists. In the case of a double slit experiment, if you change the environment by adding a detector to the slit, the quantum particle is converted into a classical particle. Similarly, when radiation inside a black body is in equilibrium with its surroundings, the emission and absorption of radiation take place as discrete packets of energy known as "quanta of energy" rather than as an electromagnetic wave. When electromagnetic radiation interacts with a metal plate, it it is converted into a beam of photons that is absorbed by the metal plate as was discussed above in section about the photoelectric effect. In Crompton scattering, the interaction between the X-rays and free electrons induces the X-rays to behave as particles, which are called X-ray photons. As a result of which the electrons rebound and X-ray photon energy is altered. For the Davisson and Germer experiment, the interaction between the free electrons and the atoms of crystals induce the electrons to behave as a wave, and thus diffraction and interference are observed. Any act of measurement to determine the properties of a quantum system will lead to interaction between the device and a quantum system. This interaction changes the properties of a quantum system in a way that it becomes impossible to predict the exact behavior of the quantum system. For example, whether an electron behaves as a wave or a particle depends on its interaction with the environment that tends to manifest "particle-like" or "wave-like" properties. In classical physics, by knowing precisely the initial position and momentum of a particle, the future position and momentum of the particle can be predicted. The interaction between the device and a system does not changes the properties of the system. Quantum theory has shown that it is impossible to predict the exact behavior of a quantum system and that the interaction between the environment and a quantum system brings out only a certain behavior or properties of the system. Let us now ask a simple question. What is the reality of an electron? Is it a particle or a wave? Is it both at the same time? Does it come into existence due to its interaction with the environment? From all the above experiments that were discussed in this chapter, at a quantum level, a quantum system does not have individual properties (such as wave or particle) by itself. It is the interaction with the environment that manifests a certain behavior. Thus, an electron can be both wave and particle at the same time, or it can continue to transition between wave and particle behavior. In quantum mechanics, the inability to observe how an electron transitions from a wave-particle system to a particle or wave system has given rise to many interpretations of quantum mechanics. Among all these interpretations, the oldest is the Copenhagen interpretation, and it is most commonly used. According to this interpretation, it is the act of measurement

FIGURE 2.10   This figure is an analogy of the wave-particle duality and illustrates a bunny and a duck (optical illusion image) in a possible dual state. Sometimes a bunny and sometimes a duck is observed. It constantly shifts its perspective from one to the other. (The image is taken from Wikimedia Commons, Credit: J. Jastrow.)

that manifests certain properties of the quantum system. Something concrete occurs during the act of measurement that leads to the collapse of the quantum wave function. Before the measurement, an electron is both a wave and a particle, and after the measurement, it manifests itself as a wave or a particle. However, the debate over this topic has continued up till today. In the 1970s and 80's experiments were performed to resolve this issue with the confirmation of the violation of Bell's inequality. Even though the Copenhagen interpretation turned out to be correct, the fundamental indeterministic behavior of quantum mechanical systems seems to surprise us about this fundamental reality at a deeper level. Some scientists tend to stick to the mathematical formalism of quantum mechanics. Others continue to search for answers and believe one day all such issues will be resolved with a complete theory (Figure 2.10).

3 Formalism

3.1 INTRODUCTION

In the last two chapters, we discussed the early ideas that led to the formulation of the theory of quantum mechanics. In this chapter, we discuss the mathematical formalism of the theory of quantum mechanics and present the differences between the theory of classical mechanics and quantum mechanics.

In classical mechanics, the very first step toward developing a theoretical model of any mechanical system is to identify its dynamical variables. Generally, dynamical variables are any set of variables (such as position, velocity, momentum, acceleration and energy, etc.) that can describe completely the configuration of a mechanical system and which will subject to change under the action of the forces. Therefore, after identifying such variables, and based upon the number of forces acting on the system and their initial conditions, the variables are specified as a function of time and formalized as relationships in the form of differential equations of motion or dynamical equations. These differential equations are then solved, often by approximate analytical and computational methods, to obtain an often complex set of information about the state of the variables in question. From these functions of the dynamical variables, information about the state of the system can be obtained for any future time.

A quantum mechanical approach is quite similar to a classical mechanics approach, in that a theoretical model is built by describing a physical system, identifying its dynamical variables, formalizing the dynamical equation in the form of differential equation for the wave function of the system, and solving such equation to determine the wave function. Information about the quantum mechanical system at a specific future time can be derived from the wave function. However, unlike classical mechanics, where all the dynamical variables of a system can be accurately measured, in quantum mechanics, all the dynamical variables of a system cannot be accurately measured simultaneously even with the most accurate devices. This indeterminacy associated with the quantum mechanical variables is due to the very nature of the universe itself. The mathematics that is used to account for the indeterminacy of quantum mechanics renders the mathematical approach of quantum mechanics distinct from that of classical mechanics. Thus, the main difference between these two approaches is the mathematical approach by which the dynamical variables are defined, the connection between dynamical variables, the wave functions and indeterminacy. Such an approach will be described in the sections that follow.

3.2 DESCRIBING A QUANTUM MECHANICAL SYSTEM: BASIC POSTULATES OF THE MODEL

3.2.1 Postulate 1: Defining an Observable

Consider a spin-1/2 system, an electron that exhibits magnetic moment that is directly related to its spin, which is also called its intrinsic angular momentum. Mathematically, the magnetic moment of an electron is:

$$\hat{\mu} = -\left(\frac{e}{m_e c}\right)\hat{S} \tag{3.1}$$

where $\hat{\mu} =$ the magnetic moment of electron, $e =$ electronic charge, $m_e =$ mass of the electron, and $c =$ the velocity of light. $\hat{S}$ (S with a hat on it) is defined as the spin or intrinsic angular momentum of the electron. Spin ($\hat{S}$) is an observable that can be measured experimentally. In quantum mechanics, any dynamical variable that is an observable that can be measured experimentally is described as a linear Hermitian operator. An operator is represented with a hat on the symbol to distinguish it from a dynamical variable, described as a function of time in classical mechanics. Further discussion about operators will be postponed until Section 3.3. The spin of an electron has components along x-, y- and z-axes, and as operators, these components of spin are represented as $\hat{S}_x, \hat{S}_y, \hat{S}_z$.

3.2.2 INDETERMINACY ASSOCIATED WITH THE MEASUREMENT OF AN OBSERVABLE

For simplicity, consider that only the Z-component of the electron's spin, $\hat{S}_z$ is measured. To do this, a sample consisting of many electrons is collected, and a spin measurement device is employed to measure the spins of the sample. Why are measurements performed on a sample of electrons instead of a single electron? We need a sample to be statistically correct; the same value of the spin must be measured from the whole sample of electrons. When measurements are performed on the whole sample consisting of N number of electrons, to measure the Z-component of the spin of the electrons, some of the electrons yield spin up $\left(+\frac{\hbar}{2}\right)$ values, some of them yield spin down $\left(-\frac{\hbar}{2}\right)$ values, and $\hbar$ is Planck's constant. This implies that the measurement of the Z-component does not yield a definite value. Thus, the spin of an electron has indeterminacy associated with it. This is unlike any classical system where the outcome of a measurement for any dynamical variable is always definite. Classically, for such a sample of electrons prepared under identical conditions, measurement outcome would yield only one value of the spin from all the electrons. This special feature of a quantum mechanical system requires a different mathematical approach than that of classical mechanics. In the section that follows, the concept of states and indeterminacy in the outcome of a measurement is described.

3.2.3 POSTULATE 2: DEFINING THE STATES ASSOCIATED WITH AN OBSERVABLE

In quantum mechanics, it is not only necessary to define an observable, but it is also necessary to define the states associated with it. What is meant by a state of an observable? A quantum state of an observable is simply related to the different values of an observable that can be observed experimentally. As discussed previously, when the spin of an electron is measured for a sample, two different values are obtained. These values depend upon which state the electron is in before the measurement is performed. Any observable of a quantum mechanical system can exist in two types of states of an observable, either eigenstate or a linear superposition of eigenstates or simply referred to as "superposition of states."

The two spin values will be associated with two spin eigenstates and simply referred to as "states." The spin-up and spin-down states of the Z-component of the spin are represented by two vectors in a vector space, called kets, and denoted by the following symbol:

$$\text{Spin up state along } z\text{-axis} = |{\uparrow}\rangle_z \tag{3.2}$$

$$\text{Spin down state along } z\text{-axis} = |{\downarrow}\rangle_z \tag{3.3}$$

Space is simply the dimension in which objects and events have direction and position. Similarly, a vector space includes dimensions in which vectors are the objects that reside in and follow the rules of linear algebra. The states of an observable in quantum mechanics are simply defined as vectors in a vector space. The dimensionality of the vector space depends on the nature of the quantum mechanical system under consideration. These vectors can be in a finite or infinite dimensional space. For example, in the previously mentioned system, the spin of an electron has two degrees of freedom (spin-up and spin-down orientations). Therefore, the states of the system represented as vectors reside in a 2-dimensional (2D) vector space. This vector space is a complex vector space because the scalars by which these vectors are multiplied can be real or complex numbers. The state of the system can be defined as an eigenstate of an observable as shown:

$$|\Psi\rangle = |\uparrow\rangle_z \tag{3.4}$$

The state of the system can also be described as a linear superposition of eigenstates, given as:

$$|\Psi\rangle = c_\uparrow |\uparrow\rangle_z + c_\downarrow |\downarrow\rangle_z \tag{3.5}$$

where $c_\uparrow$ and $c_\downarrow$ are complex numbers; the meaning of these coefficients is discussed later.

3.2.4 POSTULATE 3: MEASUREMENT OF AN OBSERVABLE

When a system is in a linear superposition of states of an observable, the number of different values of an observable observed in an experiment is equal to the number of states that a quantum mechanical system can jump into upon measurement of an observable. Thus, in a quantum mechanical system, a sample consisting of N electrons where each electron is prepared in a linear superposition of states (Equation 3.5) before measurement, the outcome of measurement of spin on such a system would yield two different values. Instead of observing a single value of the spin, two different values would be observed. However, when a measurement is performed on a quantum mechanical system that is prepared in an eigenstate (Equation 3.4) of an observable before measurement, the outcome of the measurements will yield the same value of the observable. For example, if the electrons of the sample are in a spin-up eigenstate of the Z-component of the spin before measurement, then only spin-up value will be measured from all the electrons of the sample. There is no indeterminacy in the value of the Z-component of the spin. Now, if X-component of the spin is measured for the electrons in spin-up eigenstate of the Z-component of the spin, then two different values of the X-component of the spin will be measured from the sample. This time indeterminacy is in the X-component of the spin. Therefore, if indeterminacy is removed from one of the components, then it arises in the other component. Thus, all spin components cannot be simultaneously measured with accuracy (Figure 3.1). Classically, all spin components will give definite values if measured simultaneously. Quantum mechanically, all observables of a system cannot be measured accurately simultaneously. Indeterminacy is intrinsic to quantum mechanical systems, and this is what is distinctive about quantum mechanics compared to classical mechanics. Why do quantum mechanical systems have indeterminacy? What causes such indeterminacy? The answers to these questions are not known yet, but what we do know is that the basic nature of a quantum mechanical system is very distinct from that of a classical system. When systems are indeterministic, we resort to probability. Thus, mathematical formalism of quantum mechanics involves probability into its structure. A more detailed mathematical approach of quantum mechanics is discussed in the sections that follow.

Conceptual Question 1: *Consider a quantum mechanical system that is comprised of N identical particles prepared in spin-up state of the Z-component of the spin. When measurements of the Z-component are performed, how many different values of the spin will be observed? Why? Explain.*

3.3 MATHEMATICAL FOUNDATION

3.3.1 STATE VECTOR

The state vectors defined previously can also be called ket vectors or kets. The overall spin state of an electron can be expressed as one of the ket of the Z-component of the spin operator, or as linear

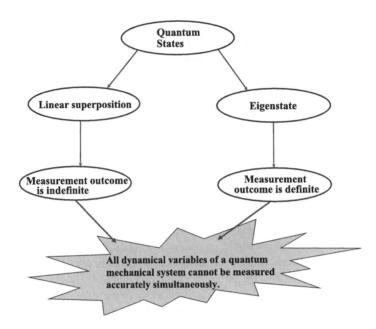

FIGURE 3.1   A schematic showing different quantum states.

superposition of the two or more kets as shown:

$$|\Psi\rangle = |\uparrow\rangle_z \tag{3.4}$$

$$|\Psi\rangle = c_\uparrow |\uparrow\rangle_z + c_\downarrow |\downarrow\rangle_z \tag{3.5}$$

where $c_\uparrow$ and $c_\downarrow$ are complex numbers. If we have a ket $|x\rangle$ corresponding to an observable $\hat{x}$, that can take on all values in a certain range ("x" is continuous variable), the state vector can be expressed as an integral:

$$|\Psi\rangle = \int_{-\infty}^{\infty} |x\rangle\langle x|\Psi\rangle \, dx \tag{3.6}$$

We will show later how to obtain the previous expression. For simplicity, the focus will be limited to discrete variables, such as spin vectors corresponding to 2D vector space (discrete variables) to better understand the mathematical foundation. Later, more complex, generalized systems will be explored.

3.3.2   KET AND BRA SPACE

The 2D vector space that was defined previously can also be called ket space. However, for every ket space, there also exists a bra space comprising of vectors that are complex conjugates of the same vectors called bra vectors or bras, and such a vector space is also called "dual space" or bra space. This implies that for every ket ($|\uparrow\rangle_z$) there exists a corresponding bra ($\langle\uparrow|_z$) which is simply a complex conjugate of the ket.

$$(|\uparrow\rangle_z)^* = \langle\uparrow|_z \tag{3.7}$$

$$(|\Psi\rangle)^* = \langle\Psi| \tag{3.8}$$

The spin-up and spin-down states are also called eigenkets of the spin operator $\hat{S}_z$.

3.3.3 Multiplication Rules for Ket and Bra

3.3.3.1 Inner Product

The product of a ket and a bra in general is a complex number as shown as:

$$(\langle\psi|)(|\varphi\rangle) = \langle\psi|\varphi\rangle = \text{Complex number} \tag{3.9}$$

This product of ket and bra where bra is on the left side and ket is on the right side is defined as an inner product.

The inner product of a ket and a bra of itself follows the following inequality or norm:

$$\langle\varphi|\varphi\rangle \geq 0 \tag{3.10}$$

where the equality sign holds only if $|\varphi\rangle$ is a null ket.

3.3.3.2 Outer Product

The product of ket and bra with bra on the right and ket on the left is called an outer product and is shown as:

$$(|\psi\rangle)(\langle\psi|) = |\psi\rangle\langle\psi| = \text{operator} \tag{3.11}$$

The outer product is fundamentally different than the inner product in the sense that the outer product is not simply a number but rather an operator. Operators will be discussed a little later, but for the time being, the outer product can be considered not to be a complex number.

3.3.3.3 Orthogonality

Two kets $|\varphi\rangle$ and $|\psi\rangle$ are said to be orthogonal if their inner product is zero

$$\langle\varphi|\psi\rangle = 0 \tag{3.12}$$

This implies that its complex conjugate is also orthogonal:

$$\langle\psi|\varphi\rangle = 0 \tag{3.13}$$

In the previous example, spin-up and spin-down kets are orthogonal because their inner product is zero.

$$\langle\uparrow|\downarrow\rangle_z = 0 \tag{3.14}$$

3.3.3.4 Normalization

Any ket can be normalized by using the following relationship:

Consider a ket $|\psi\rangle$ and its normalized ket $|\tilde{\psi}\rangle$ can be written as:

$$|\tilde{\psi}\rangle = \left(\frac{1}{\sqrt{\langle\psi|\psi\rangle}}\right)|\psi\rangle \tag{3.15}$$

with the property

$$\langle\tilde{\psi}|\tilde{\psi}\rangle = 1 \tag{3.16}$$

And $\sqrt{\langle\psi|\psi\rangle}$ is known as the norm of the ket $|\psi\rangle$. The previous relation is called *normalization* of the state vector.

Spin-up and spin-down kets are also normalized, and their inner product with themselves is unity.

$$\langle \uparrow | \uparrow \rangle_z = \langle \downarrow | \downarrow \rangle_z = 1 \tag{3.17}$$

As discussed previously, they are also orthogonal and normalized, which implies that the eigen-kets are orthonormal. Such vectors are also called as base kets of a 2D vector space. Thus, eigenkets of S_z can be used as base kets in a vector space in much the same way as a set of mutually orthogonal unit vectors are used as base vectors in Euclidean space.

Why are these kets orthogonal, and what does this imply?

If the outcome of a measurement of an observable is certain to yield different results, then the states associated with each result are distinguishable and hence orthogonal. In the previous example, the measurement of spin leads to two different values, and therefore, the states associated with each value are distinguishable and hence orthogonal. Any two eigenvectors of a Hermitian operator corresponding to two different outcomes of a measurement are always orthogonal.

EXAMPLE 3.1

Consider the following state vectors, $|\alpha\rangle = \frac{1}{\sqrt{2}}|0\rangle - \frac{i}{\sqrt{2}}|1\rangle$, and $|\beta\rangle = \frac{1}{\sqrt{2}}|0\rangle + \frac{i}{\sqrt{2}}|1\rangle$, where $|0\rangle$ and $|1\rangle$ are orthonormal base vectors of a 2D Hilbert space. Determine the complex conjugate of these vectors. Also, compute their inner products.

$$(|\alpha\rangle)^* = \langle \alpha | = \frac{1}{\sqrt{2}}\langle 0| + \frac{i}{\sqrt{2}}\langle 1| \tag{3.18}$$

$$(|\beta\rangle)^* = \langle \beta | = \frac{1}{\sqrt{2}}\langle 0| - \frac{i}{\sqrt{2}}\langle 1| \tag{3.19}$$

$$\langle \alpha | \alpha \rangle = \frac{1}{2} + \frac{1}{2} = 1 \tag{3.20}$$

$$\langle \beta | \beta \rangle = \frac{1}{2} + \frac{1}{2} = 1 \tag{3.21}$$

EXAMPLE 3.2

Consider a state vector, $|\Psi\rangle = a|\uparrow\rangle + b|\downarrow\rangle$, where a and b are complex numbers. Determine its normalized state vector.

Using Equation (3.15), normalized state vector can be expressed as:

$$|\tilde{\psi}\rangle = \frac{a}{\sqrt{a^2+b^2}}|\uparrow\rangle + \frac{b}{\sqrt{a^2+b^2}}|\downarrow\rangle \tag{3.22}$$

Check whether it is normalized or not.

$$\langle \tilde{\psi} | \tilde{\psi} \rangle = \left(\frac{a^*}{\sqrt{a^2+b^2}}\langle \uparrow | + \frac{b^*}{\sqrt{a^2+b^2}}\langle \downarrow | \right) * \left(\frac{a}{\sqrt{a^2+b^2}}|\uparrow\rangle + \frac{b}{\sqrt{a^2+b^2}}|\downarrow\rangle \right) \tag{3.23}$$

$$\langle \tilde{\psi} | \tilde{\psi} \rangle = \frac{a^2}{a^2 + b^2} + \frac{b^2}{a^2 + b^2} = 1 \tag{3.24}$$

3.3.4 OPERATORS

An operator is described as an instruction for transforming a ket into another ket. For example, a ket $|\psi\rangle$ can be transformed into another ket by an operator $\hat{T}$, with a hat on symbol just to distinguish it from notation for a physical variable in classical mechanics.

$$\hat{T}|\psi\rangle = |\bar{\psi}\rangle \tag{3.25}$$

If

$$\hat{T}|\psi\rangle = \varepsilon|\psi\rangle \tag{3.26}$$

where ε is a real number, then $\hat{T}$ is identified as a Hermitian operator, ε is identified as an eigenvalue, and $|\psi\rangle$ is identified as an eigenket or eigenstate of an operator $\hat{T}$. Such an operator is equal to the complex conjugate of itself and thus satisfies, $\hat{T} = \hat{T}^*$.

All dynamical variables that are observables are represented by Hermitian operators, and these operators do not have any physical significance. They are mathematical entities, and they are used to calculate averages of the observables.

An operator is called anti-Hermitian (or skew Hermitian) when a complex conjugate is negative of the operator, shown as:

$$\hat{M} = -\hat{M}^* \tag{3.27}$$

The eigenvalues of such operator are complex numbers and not real.

3.3.5 BASE KETS

Consider a Hermitian operator S_z, which measures the spin or intrinsic angular momentum of the electron along the z-axis and can be described as follows:

$$\hat{S}_z|\uparrow\rangle_z = +\frac{\hbar}{2}|\uparrow\rangle_z \tag{3.28}$$

$$\hat{S}_z|\downarrow\rangle_z = -\frac{\hbar}{2}|\downarrow\rangle_z \tag{3.29}$$

The eigenvalues $\pm\frac{\hbar}{2}$ are real because S_z is an observable. The eigenvalues of any variable that is an observable must be a real number because in an experiment, the measured values of any observable can only be a real number.

As discussed previously, operators can be written as an outer product of ket and bra. S_z can be written as the outer product of spin-up and spin-down kets and bras (Figure 3.2).

$$\hat{S}_z = \frac{\hbar}{2}\left[(|\uparrow\rangle\langle\uparrow|_z) - (|\downarrow\rangle\langle\downarrow|_z)\right] \tag{3.30}$$

Since $\hat{S}_z$ is a Hermitian operator, its eigenvalues are real, and eigenkets are orthonormal. Can these eigenkets be used as base kets?

3.3.6 COMPLETENESS

For these eigenkets to be used as base kets, they must form a complete set. Mathematically, this implies that the outer product of all of the eigenkets by themselves must add up to unity. The unity on the left hand side can be called as identity operator. Since outer product of an eigenket by itself is an operator. Thus, all these operators of all the eigenkets combined together must form an identity operator. The mathematical equation describing this statement is shown below,

$$1 = |\uparrow\rangle_z\langle\uparrow| + |\downarrow\rangle_z\langle\downarrow| \tag{3.31}$$

The previous equation is identified as the *completeness relation* or *closure*.

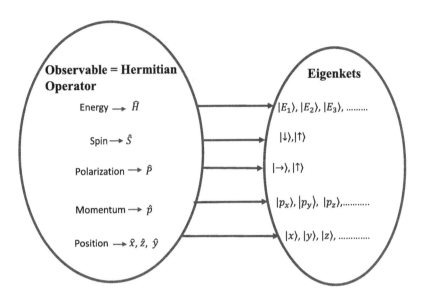

FIGURE 3.2   Figure illustrates several observables as Hermitian operators and their corresponding eigenkets.

In Euclidean space, base vectors have unit magnitude. Similarly, in vector space, the inner product of a base ket with itself is unitary, and the inner product of a base ket with a complex conjugate of any other base ket is zero (Equation 3.13), and they are called orthonormal base kets. Thus, these eigenkets can be used as base kets and form a complete set in 2D vector space. Any vectors that are orthogonal and form a complete set in a vector space can be used as base vectors. The main advantage of these base vectors is that any arbitrary vector in a vector space can be expanded in terms of the base vectors of the same vector space.

Further explanation concerning the completeness relation will be outlined later. Returning to Equation (3.5),

$$|\Psi\rangle = c_\uparrow |\uparrow\rangle_z + c_\downarrow \langle\downarrow|_z$$

To obtain the previous equation from the completeness relation Equation (3.31), multiply Equation (3.31) with $|\Psi\rangle$

$$|\Psi\rangle = |\uparrow\rangle_z \langle\uparrow|\Psi\rangle_z + |\downarrow\rangle_z \langle\downarrow|\Psi\rangle_z \tag{3.32}$$

which gives

$$|\Psi\rangle = c_\uparrow |\uparrow\rangle_z + c_\downarrow |\downarrow\rangle_z \tag{3.33}$$

where

$$\langle\uparrow|\Psi\rangle_z = c_\uparrow \tag{3.34}$$

$$\langle\downarrow|\Psi\rangle_z = c_\downarrow \tag{3.35}$$

Therefore, $c_\uparrow$ is simply an inner product of the spin-up bra and the state vector and is defined as a coefficient that is a complex number. This exercise describes the mathematical form of these coefficients.

TABLE 3.1

A Comparison between Euclidean Space and Vector Space

Complex Vector Space	Euclidean Space
1. Base vectors are orthonormal. Such vectors reside in two or more (infinite) dimensional vector space.	Base vectors, also known as unit vectors for 2D (dimensional) or 3D Cartesian coordinate system, are orthonormal. Such vectors could be two or more (finite) dimensional space.
2. Any arbitrary vector can be represented using base vectors.	Any arbitrary vector can be represented using base vectors.
3. Base vectors satisfy the closure property.	No closure property.

So far, it has been demonstrated that the base vectors of a vector space have similar properties to those of base vectors in Euclidean space. A comparison is shown in the following Table 3.1.

However, what is the physical meaning of coefficients $c_\uparrow$ & $c_\downarrow$?

By multiplying bra of the previous state vector (Equation 3.5) with its ket, we get:

$$\langle\Psi|* (|\Psi\rangle = |\uparrow\rangle_z\langle\uparrow|\Psi\rangle + |\downarrow\rangle_z\langle\downarrow|\Psi\rangle) \tag{3.36}$$

The notation can be simplified as:

$$|\uparrow\rangle_z = |\uparrow\rangle \tag{3.37}$$

$$|\downarrow\rangle_z = |\downarrow\rangle \tag{3.38}$$

and the following equation is obtained:

$$\langle\Psi|\Psi\rangle = \langle\Psi|\uparrow\rangle\langle\uparrow|\Psi\rangle + \langle\Psi|\downarrow\rangle\langle\downarrow|\Psi\rangle \tag{3.39}$$

which can also be written as:

$$\langle\Psi|\Psi\rangle = c_\uparrow^* c_\uparrow + c_\downarrow^* c_\downarrow \tag{3.40}$$

where

$$c_\uparrow = \langle\uparrow|\Psi\rangle \tag{3.41}$$

$$c_\downarrow = \langle\downarrow|\Psi\rangle \tag{3.42}$$

Since the state vector must be a normalized state, this implies that:

$$\langle\Psi|\Psi\rangle = 1 \tag{3.43}$$

Therefore,

$$1 = c_\uparrow^* c_\uparrow + c_\downarrow^* c_\downarrow \tag{3.44}$$

or,

$$1 = |c_\uparrow|^2 + |c_\downarrow|^2 \tag{3.45}$$

Examining the physical meaning of these coefficients is the next step. So far, we have not discussed probability due to indeterminism of the quantum mechanical system. Let us now see whether these coefficients can account for the probability. It is certain that these coefficients are complex numbers, and the absolute square of these numbers must add up to unity in accordance with the probability law. Another clue is that each coefficient corresponds to a spin-up or spin-down state, which implies that these numbers must be revealing something about the states of the system. According to quantum mechanics, when a measurement of an observable is performed on a quantum mechanical system, the system collapses into one of the eigenstates of the observable, and the absolute square of these coefficients is simply the probability that a measurement of an observable would yield a certain eigenvalue of an observable due to the system collapsing into a particular eigenstate. For example, when a measurement is performed to measure the spin of an electron, then the probability that a measurement of spin would yield spin-up value due to an electron state collapsing into a spin-up state is given by the absolute square of the coefficient $|c_\uparrow|^2$. Therefore, the absolute square of coefficients can be described as the following:

$|c_\uparrow|^2 =$ *is the probability that a measurement of spin would yield spin-up value due to the electron's state collapsing into the spin-up level*

$|c_\downarrow|^2 =$ *is the probability that a measurement of spin would yield spin-down value due to the electron's state collapsing into the spin-down level*

(Note: The spins are measured along the Z-axis.)

Due to the closure property (completeness), the sum of two probabilities must add up to unity (Equation 3.43) so that the probability is preserved. Another interesting observation is that upon measurement of spin, the electron collapses into one of the eigenstate of the spin where it will remain in that state forever. Any post measurements would yield the same eigenvalue again and again. This transition of the state of spin from a linear superposition of state to an eigenstate is also known as a collapse of the state of the spin. The measurement causes collapse of the super-position state.

Consider a sample of N electrons where all the electrons are prepared in an identical super-position of states (Equation 3.5). Measurement of Z-component of the spin is performed on this sample. Upon measurement, half of the electrons ($N/2$) yield spin-up value, and the remaining half yield spin-down value. This suggest that: $|c_\uparrow|^2 = (1/2)$ and $|c_\downarrow|^2 = (1/2)$ and their initial state vector was:

$$|\Psi\rangle = c_\uparrow |\uparrow\rangle_z + c_\downarrow |\downarrow\rangle_z = \frac{1}{\sqrt{2}} |\uparrow\rangle_z + \frac{1}{\sqrt{2}} |\downarrow\rangle_z \qquad (3.46)$$

and after the measurement, due to collapse of the state vector, half of the electrons are now in spin-up state $|\uparrow\rangle_z$ and the remaining half are in spin-down state $|\downarrow\rangle_z$.

However, how are such coefficients determined theoretically?

This question will be addressed in the next section of the chapter.

Conceptual Question 2: *Consider a particle, a two-state system whose total energy eigen-vectors are $|0\rangle$ and $|1\rangle$ and corresponding energy eigenvalues are E_0 and E_1. When measure-ment is performed for the total energy of the system on a sample of N particles prepared in a superposition of states of these energy eigenvectors, half of the particles ($N/2$) measured energy is E_0, and the remaining half of the particles measured energy is E_1. Using this information, construct the initial state vector of the system. Instead of half of the particles, if 75% of the particles measured energy is E_0, and the remaining 25% of the particles mea-sured energy is E_1, construct their initial state vector (Figure 3.3).*

3.3.6.1 Discrete Spectrum

In the previous section on operators, the eigenvalues of operator $\hat{S}_z$ are distinct, and each eigenvalue is associated with a unique eigenvector of the operator. These eigenvalues are called nondegenerate

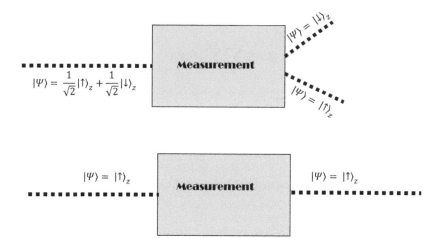

FIGURE 3.3   The above schematic shows the collapse of the state vector of a spin Z state when a measurement is performed. If N particles are prepared in the above superposition state, then the state of half of them collapses into a spin-up state, and the state of the other half collapses into a spin-down state when a measurement of the Z-component of the spin of the particle is performed However, when N particles are prepared in spin-up eigenstate, then upon measurement, they remain in the same state.

eigenvalues. We will now move to generalization of a two-level system to an infinite level system. For any operator $\hat{A}$, whose eigenvalues $\{a_1, a_2, a_3, \dots\}$ are distinct, each associated with a unique eigenvector $\{|A_1\rangle, |A_2\rangle, |A_3\rangle, \dots\}$, such eigenvalues form a discrete spectrum of nondegenerate eigenvalues. All these eigenvectors form a complete set of orthonormal vectors and can be used as base vectors. Such that:

$$1 = \sum_n |A_n\rangle\langle A_n| \tag{3.47}$$

Thus, the state vector of any system can be described as a linear superposition of the eigenvectors as:

$$|\Psi\rangle = \sum_n |A_n\rangle\langle A_n|\Psi| \tag{3.48}$$

$$\langle A_n|\Psi\rangle = c_n$$

As the inner product is a complex number, thus the state vector is:

$$|\Psi\rangle = \sum_n c_n|A_n\rangle \tag{3.49}$$

Thus, the probability of measuring value a_n when A is measured will be:

$$P(a_n) = |c_n|^2 = |\langle A_n|\Psi\rangle|^2 \tag{3.50}$$

However, this is not always true. For any operator $\hat{A}$, two or more eigenvectors can be associated with the same eigenvalue. Such eigenvalues are called as degenerate eigenvalues and can be described as:

$$\hat{A}|A_n^i\rangle = a_n|A_n^i\rangle, \quad i = 1, 2, 3 \dots d_n \tag{3.51}$$

where $d_n =$ the degree of degeneracy, and eigenvectors are orthonormal.

The state vector of the system in terms of such eigenvectors can be written as:

$$|\Psi\rangle = \sum_n \sum_i^{d_n} c_n^i |A_n^i\rangle \tag{3.52}$$

In this case, the probability becomes:

$$P(a_n) = \sum_{i=1}^{d_n} |c_n^i|^2 = \sum_{i=1}^{d_n} |\langle A_n^i | \Psi \rangle|^2 \tag{3.53}$$

Degenerate states are more complicated, and for simplicity, only nondegenerate states will be considered.

3.3.6.2   Continuous Spectrum

If the eigenvalues of the operator $\hat{A}$ form a set of nondegenerate continuous values, then the spectrum is called a continuous spectrum. The eigenvectors of $\hat{A}$ can be described as:

$$\hat{A}|A_\alpha\rangle = \alpha|A_\alpha\rangle \tag{3.54}$$

where $\alpha = $ a continuous variable and eigenvalue. If eigenvectors form a continuous set of ortho-normal eigenvectors such that,

$$1 = \int d\alpha |A_\alpha\rangle\langle A_\alpha| \tag{3.55}$$

The state vector can now be written as:

$$|\Psi\rangle = \int d\alpha |A_\alpha\rangle\langle A_\alpha|\Psi\rangle \tag{3.56}$$

The previous inner product will be a set of continuous complex numbers, and thus, the inner product becomes a function. Therefore,

$$\langle A_\alpha|\Psi\rangle = c(\alpha) \tag{3.57}$$

The previous coefficient can also be called as an eigenfunction.

Since α is a continuous variable, the probability of measuring an eigenvalue included between α and $\alpha + d\alpha$ will be:

$$dP(\alpha) = \rho(\alpha)d\alpha \tag{3.58}$$

where

$$\rho(\alpha) = |c(\alpha)|^2 = |\langle A_\alpha|\Psi\rangle|^2 \tag{3.59}$$

is called the probability density. More details on this concept will be discussed in the section that follows.

PROBLEMS

3.1   Consider the following state vectors, $|\Psi\rangle = C|\uparrow\rangle + \sin(\theta)|\downarrow\rangle$, $|\varphi\rangle = Ce^{-i\gamma}|\uparrow\rangle + \cos(\theta)|\downarrow\rangle$. Find the coefficient C by using normalization condition of the state vector.

3.2 Show that the following states $|\Psi\rangle = \frac{1}{\sqrt{2}}(|0\rangle + |1\rangle)$, $|\varphi\rangle = \frac{1}{\sqrt{2}}(|0\rangle - |1\rangle)$ are orthogonal. Use these states as base states and show that they satisfy completeness identity.

3.3 Consider that $\hat{A}$ and $\hat{B}$ are two Hermitian operators. Show that the sum of these operators is also a Hermitian operator. Also, show that the product of these operators is Hermitian only when $\hat{A}\hat{B} - \hat{B}\hat{A} = 0$.

3.4 Find the eigenfunctions and eigenvalues of the Hermitian operator, $\hat{P} = i\hbar \, (d/d\phi)$. Here, ϕ is the usual polar coordinate in two dimensions.

3.4 WAVE FUNCTION

In this section, questions concerning the wave function will be addressed. For example, what is a wave function? When and why it is used, and what information about the quantum mechanical system can be derived from it? To answer these questions, the focus is now on new concepts such as eigenfunctions and its relationship with eigenvectors, Hilbert space, complex functions and some new notations.

3.4.1 Function Space and Hilbert Space

So far, we have discussed that the state of a quantum system is described by vectors in vector spaces and observables as operators. The vectors in the vector space are a class of simple functions, the polynomials. In this section, we continue along similar lines. However, we now introduce vector space whose elements are complex valued functions of a real variable, defined on a close interval, and which are square integrable. This space is called L_2 by mathematicians and is a vector space. To distinguish it from previous vector space, we shall name it function space. The infinite-dimensional function space is known as Hilbert space. Generally, quantum mechanical states of any system are described in Hilbert space.

Consider a position observable x, which is a continuous variable, and its operator as $\hat{x}$, and $|x_i\rangle$ are its eigenvectors. Such that,

$$\hat{x}|x_i\rangle = x_i|x_i\rangle \tag{3.60}$$

where x_i = the eigenvalue of the position variable. The eigenvectors form a complete orthonormal basis set such that,

$$1 = \int dx|x\rangle\langle x| \tag{3.61}$$

The state vector can be written as:

$$|\Psi\rangle = \int dx|x\rangle\langle x|\Psi\rangle \tag{3.62}$$

$$\langle x'|\Psi\rangle = \int dx\langle x'|x\rangle\langle x|\Psi\rangle = \int dx\delta(x - x')\langle x|\Psi\rangle \tag{3.63}$$

where $\delta(x - x')$ is called Dirac delta function such that $\int_{-\infty}^{\infty} f(x)\delta(x - x')dx = f(x')$

Thus,

$$\langle x'|\Psi\rangle = \Psi(x') \tag{3.64}$$

or,

$$\langle x | \Psi \rangle = \Psi(x) \tag{3.65}$$

The previous inner product is a function of a continuous variable and is also called as wave function. To further understand the meaning of the wave function, consider H as the energy operator of a system, also called as the Hamiltonian. Such that,

$$\hat{H} | E_n \rangle = E_n | E_n \rangle \tag{3.66}$$

where $| E_n \rangle$ are energy eigenvectors of the system that form a discrete complete orthonormal set. Such that,

$$1 = \sum_n | E_n \rangle \langle E_n | \tag{3.67}$$

The state vector can be written as:

$$| \Psi \rangle = \sum_n | E_n \rangle | \langle E_n | \Psi \rangle \tag{3.68}$$

To write the previous state vector in terms of the wave function, we take the inner product with the base vectors of the continuous position variable, thus

$$\langle x | \Psi \rangle = \sum_n \langle x | E_n \rangle \langle E_n | \Psi \rangle \tag{3.69}$$

where

$$\langle x | E_n \rangle = u_{E_n}(x) \tag{3.70}$$

is called as energy eigenfunction, and

$$\langle E_n | \Psi \rangle = c_n \tag{3.71}$$

is an expansion coefficient (complex number), thus,

$$\Psi(x) = \sum_n c_n u_{E_n}(x) \tag{3.72}$$

The previous is the wave function of a system, which is simply a linear superposition of the energy eigenfunctions of a system.

To further illustrate the use of the eigenfunctions, consider a particle whose energy states are described in a 2-dimensional function space. The particle has two energy levels, the energy operator of the system is the Hamiltonian operator, $\hat{H}$, and its energy eigenfunctions are:

$$\hat{H} \, u_{E_1}(x) = E_1 \, u_{E_1}(x) \tag{3.73}$$

$$\hat{H} \, u_{E_2}(x) = E_2 \, u_{E_2}(x) \tag{3.74}$$

The state vector of the system is described by the following:

$$|\Psi\rangle = c_1|E_1\rangle + c_2|E_2\rangle \tag{3.75}$$

and the wave function is given as:

$$\Psi(x) = c_1 \, u_{E_1}(x) + c_2 \, u_{E_2}(x) \tag{3.76}$$

The coefficients c_1 and c_2 have similar meaning as that of the spin state coefficients $c_\downarrow$ & $c_\uparrow$. Precisely,

$|c_1|^2$ *is the probability that a measurement of energy would yield a E_1 value due to the particle's state collapsing into the energy level 1.*

$|c_2|^2$ *is the probability that a measurement of energy would yield a E_2 value due to the particle's state collapsing into the energy level 2.*

As described previously, instead of state vector, the system's state is now described by a complex function, namely wave function, which is a function of real and continuous variable 'x.' To every state vector, there exists a corresponding wave function. Thus,

$$|\Psi\rangle \Leftrightarrow \Psi(x) \tag{3.77}$$

This wave function simply could be an energy eigenfunction or a linear superposition of energy eigenfunctions. In a 3-dimensional coordinate system, wave function can be represented as:

$$|\Psi\rangle \Leftrightarrow \Psi(x,y,z) \tag{3.78}$$

You may be wondering by now why such a function is called a wave function. Does it represent a wave? The wave function does not represent any real wave such as electromagnetic wave, but the function can be visualized as a probability wave in space (having wiggles for some cases), but this visualization works mainly for a single particle as a quantum system. Instead, if we consider a bunch of particles that are interacting with each other, then this visualization breaks down. Further reasons for calling such a function a wave function become clear as we try to understand the Schrodinger equation.

3.4.2 Eigenfunctions

Previous eigenfunctions have the following properties:

3.4.2.1 Orthogonal

These functions are mutually orthogonal in the sense that:

$$\int u_{E_1}^*(x)u_{E_2}(x)dx = 0 \tag{3.79}$$

Generally, one can describe the orthogonal condition of any eigenfunctions corresponding to two different states of an observable as:

$$\int u_n^*(x)u_m(x)dx = 0 \tag{3.80}$$

3.4.2.2 Normal

These eigenfunctions must satisfy the normalization condition as shown:

$$\int u_{E_1}^*(x) u_{E_1}(x) dx = 1 \tag{3.81}$$

$$\int u_n^*(x) u_n(x) dx = 1 \tag{3.82}$$

By combining these two properties, we can say that eigenfunctions of an observable of a system satisfy *orthonormality condition* as shown:

$$\int u_n^*(x) u_m(x) dx = \delta_{mn} \tag{3.83}$$

where δ_{mn} is called as Kronecker delta function and is defined in the following way:

$$\begin{aligned} \delta_{mn} &= 1 \text{ (for } m = n) \\ &= 0 \text{ (for } m \neq n) \end{aligned} \tag{3.84}$$

3.4.2.3 Completeness

These eigenfunctions form a complete set, in the sense that any other function, $f(x)$ can be expressed as a linear combination of such functions as shown:

$$f(x) = \sum_{n=1}^{\infty} c_n u_n(x) \tag{3.85}$$

Now, if you know the form of this function $f(x)$, you can find coefficients c_n by using the orthonormality condition of the eigenfunctions as shown:

$$\int u_m(x)^* f(x) dx = \sum_{n=1}^{\infty} c_n \int u_m(x)^* u_n(x) dx = \sum_{n=1}^{\infty} c_n \delta_{mn} \tag{3.86}$$

Thus,

$$c_n = \int u_n(x)^* f(x) dx \tag{3.87}$$

As you can see in the previous equation, these coefficients depend on the form of eigenfunctions and some function $f(x)$, which means that these coefficients depend on the physical conditions of the system under study. We will study several examples in the next chapter for finding such coefficients based on the physical conditions of the system.

3.4.3 PROBABILITY DENSITY

To understand the physical meaning of the wave function, we must understand the meaning of probability density.

The probability density of a particle whose wave function is $\Psi(x)$ as follows:

$$|\Psi(x)|^2 = \text{probability density} \tag{3.88}$$

Probability density simply provides the probability of finding the particle at location "x" when a measurement of position is performed. For this reason, wave function is also known as probability function.

The probability that an electron lies between x and $x + dx$ (infinitesimal interval) would be:

$$|\Psi(x)|^2 dx \tag{3.89}$$

Also, the probability that an electron lies between location x_a and x_b (finite interval) would be:

$$P_{ab} = \int_{x_a}^{x_b} |\Psi(x)|^2 dx \tag{3.90}$$

This is also called as *Born's* statistical interpretation of the wave function. This means that by using a wave function, the probability of finding the particle at a particular location can be predicted probabilistically, which is simply a statistical information and not the exact information. Thus, quantum mechanically, the position of a particle cannot be predicted exactly before a measurement is performed.

As the particle must be somewhere, if we take the previous integral over the entire universe, the probability of finding the particle somewhere in the universe must be unity as shown:

$$1 = \int_{-\infty}^{\infty} |\Psi(x)|^2 dx \tag{3.91}$$

The previous equation is called as the normalization condition for the wave function of a system or particle.

Further, one can use the following notation to describe the probability density. (Figure 3.4).

$$\rho(x) = |\Psi(x)|^2 \tag{3.92}$$

Conceptual Question 3: *Consider Figure 3.4 that shows the probability density vs position graph of a particle. Regions a, b and c are labeled. In which region will the probability of finding the particle will be maximum and minimum?*

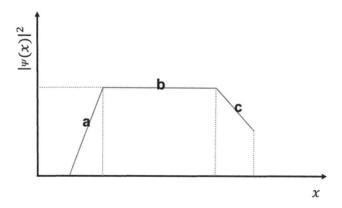

FIGURE 3.4 This figure shows probability density of a wave function.

To further understand the meaning of the wave function, consider the previous wave function:

$$\Psi(x) = c_1\, u_{E_1}(x) + c_2\, u_{E_2}(x) \tag{3.93}$$

Probability density of the previous wave function can be expressed as,

$$|\Psi(x)|^2 = |c_1|^2|u_{E_1}(x)|^2 + |c_2|^2|u_{E_2}(x)|^2 + c_1 c_2^* u_{E_1}(x)u_{E_2}(x)^* + c_1^* c_2 u_{E_1}(x)^* u_{E_2}(x) \tag{3.94}$$

The previous equation implies that

$$|\Psi(x)|^2 = P_1(x) + P_2(x) + c_1 c_2^* u_{E_1}(x)u_{E_2}(x)^* + c_2 c_1^* u_{E_2}(x)u_{E_1}(x)^* \tag{3.95}$$

The first two terms in the previous equation are the probability densities of the particle when it is in energy level 1 and energy level 2, and the last two terms are referred to as interference terms and contribute toward quantum mechanical indeterminacy for the location of the particle in space.

What is the combined meaning of the coefficients and the complex eigenfunctions?

The coefficients along with eigenfunctions yield the probability density of the location of the particle in space when the particle is in a certain energy level. As already discussed, these coefficients depend on the eigenfunctions and initial wave function of the system.

An analogy for this could be the action of locating a book in a bookshelf. If you know which row the book will be in, you can easily find the book in that particular row. For this analogy, a book takes the place of a particle. The information about the row in which the book is located is represented by the complex coefficients, and the information where the book is positioned in the row is given by the eigenfunctions. Thus, a wave function has complete information about the system. However, only probabilistic information can be extracted from the wave function.

What is the use of a wave function?

A wave function has many uses. Using wave function, not only the location of a particle can be determined but many more observables of a system can be determined. In classical mechanics, position is the most important observable. Using position observable of a system, all other observables such as velocity, momentum, energy, etc., of the system can be determined. Similarly, in quantum mechanics, wave function is the most important function. By determining a wave function, all other observables of the system can be evaluated. Wave functions are used to determine the averages or expectation values of the observables such as position, momentum, energy, etc. Wave function contains all possible information relating to the system.

3.4.4 EXPECTATION VALUE OF AN OBSERVABLE

The expectation value of an observable is the average value of an observable obtained by preparing a large number of identical systems called an ensemble of systems in the same initial state and performing measurements on them. The average value of the observable obtained by performing these measurements on an ensemble of identically prepared systems, and not the average of repeated measurements on one and the same system, is known as the expectation value of that observable. For example, to find the expectation value of position of a particle, a large number of such identical particles (sample) are prepared in the same initial state, and measurements of position are performed on such a sample. Mathematically, the expectation value of a position of a particle is described as follows:

3.4.4.1 Position

For an electron whose wave function is given by Equation (3.75), the expectation value of the position x is given as the following equation:

$$\langle x \rangle = \langle \Psi(x)|x|\Psi(x) \rangle = \int_{-\infty}^{\infty} x|\Psi(x)|^2 dx \tag{3.96}$$

$$\langle x \rangle = \int_{-\infty}^{\infty} x\,\rho(x)dx \tag{3.97}$$

3.4.4.2 Momentum

The momentum operator is given as:

$$\hat{p} = -i\hbar \frac{\partial}{\partial x} \tag{3.98}$$

Its expectation value is given by:

$$\langle p \rangle = m \frac{d\langle x \rangle}{dt} = \langle \Psi(x)|\hat{p}|\Psi(x)\rangle = -i\hbar \int_{-\infty}^{\infty} \Psi^* \frac{\partial \Psi}{\partial x} dx \tag{3.99}$$

3.4.4.3 Kinetic Energy

The kinetic energy operator is given as:

$$T = -\frac{\hbar^2}{2m} \frac{\partial^2}{\partial x^2} \tag{3.100}$$

Its expectation value is given by the following equation:

$$\langle T \rangle = \langle \Psi(x)| -\frac{\hbar^2}{2m} \frac{\partial^2}{\partial x^2}|\Psi(x)\rangle = -\frac{\hbar^2}{2m} \int_{-\infty}^{\infty} \Psi^*(x) \frac{\partial^2 \Psi(x)}{\partial x^2} dx \tag{3.101}$$

In answering the questions that were posed at the beginning of this section, the form and purpose of the wave function has been introduced.

So far, a mathematical structure in the formalism describing a quantum mechanical system has been constructed. The diagram that follows illustrates such a structure (Figure 3.5).

Why do we calculate averages of observables in quantum mechanics?

As previously discussed, quantum mechanics has indeterminacy associated with the measurements of observables. Therefore, the only way to gain valuable information from the wave function is to evaluate statistically correct values of the observables.

However, how is a wave function for any system determined? The answer is by solving the Schrodinger equation.

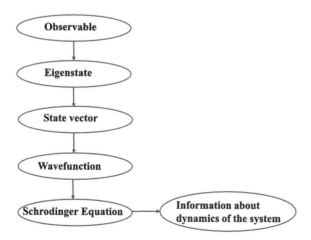

FIGURE 3.5 This schematic describes the structure in the formalism of quantum mechanics. It shows that as a first step, observables must be defined, followed by the state vector and so on.

The time dependence of a state vector is excluded so far. To describe the time-dependence of a state vector of a quantum mechanical system or simply dynamics of a quantum mechanical system, the solution to the Schrodinger's equation is required. This equation will be introduced in the section that follows.

Conceptual Question 4: *Explain the difference between a state vector in vector space and a wave function in function space.*

EXAMPLE 3.3

Consider the wave function of a particle given as:

$$\Psi(x) = \begin{cases} \dfrac{1}{A}\sqrt{x} & 0 \le x \le a \\ \\ 0, & \text{otherwise} \end{cases} \tag{3.102}$$

Find the normalization constant A, an expectation value of the position $\langle x \rangle$ of a particle, and "a" is a constant.

Using normalization condition Equation (3.91), we get:

$$1 = \int_{-\infty}^{\infty} |\Psi(x)|^2 dx = \int_{-\infty}^{0} |\Psi(x)|^2 dx + \int_{0}^{a} |\Psi(x)|^2 dx + \int_{a}^{\infty} |\Psi(x)|^2 dx \tag{3.103}$$

$$1 = \frac{1}{A^2} \int_{0}^{a} x \, dx$$

$$A = \frac{a}{\sqrt{2}} \tag{3.104}$$

$$\Psi(x) = \frac{\sqrt{2x}}{a} \tag{3.105}$$

$$\langle \hat{x} \rangle = \int_{0}^{a} x |\Psi(x)|^2 \, dx = \frac{2}{a^2} \int_{0}^{a} x^2 \, dx = \frac{2a}{3} \tag{3.106}$$

PROBLEMS

3.5 The wave function of a particle is given as: $\Psi(x) = Ae^{-\gamma x}$. Normalize the wave function and find the coefficient A. Also, find $\langle x \rangle$ and $\langle x^2 \rangle$.

3.6 The Hamiltonian operator of a three-level atom is given as: $\hat{H} = e_1|1\rangle\langle 1| + e_2|2\rangle\langle 2| - e_3|3\rangle\langle 3|$. Find the eigenvalues and eigenvectors of the Hamiltonian.

3.7 The wave function of a particle of mass "m" is the following, $\Psi(x) = Ae^{-kx^2}$, where $A =$ the normalization constant, and $k = $ a constant. You can take the integration limits from $-\infty$ to ∞. Also find $\langle x \rangle$, $\langle T \rangle$ and $\langle p \rangle$.

3.8 Mathematically derive the momentum space wave function from the state vector $|\Psi\rangle$. Show all the steps leading to such a function.

3.9 Determine the Hamiltonians of two- and five-level atoms. Describe their energy eigenvalues and eigenvectors.

3.5 DYNAMICAL PROPERTIES OF THE SYSTEM

3.5.1 TIME-DEPENDENT SCHRODINGER EQUATION

The Schrodinger equation is a dynamical equation that describes how the physical state or wave function of a quantum mechanical system changes with time. It is a deterministic equation in the sense that, given the initial conditions, it can determine the state or wave function of a system for all future times just like any equation of motion in classical mechanics. It also has a similar mathematical form as that of a classical wave equation.

The Schrodinger equation in terms of a state vector or state ket is written as:

$$ i\hbar \frac{\partial |\Psi, t\rangle}{\partial t} = \hat{H} |\Psi, t\rangle \qquad (3.107) $$

The time dependence in the ket notation is introduced as the state describe the dynamics and time-dependent changes in the system, and H is a Hamiltonian operator (Hermitian) for the total energy of the system. For a single spatial coordinate system, the Hamiltonian operator can be written as:

$$ \hat{H} = -\frac{\hbar^2}{2m}\frac{\partial^2}{\partial x^2} + \hat{V} \qquad (3.108) $$

The potential $\hat{V}$ is a Hermitian operator.

The Schrodinger equation in terms of a wave function for the previous Hamiltonian is:

$$ i\hbar \frac{\partial \Psi(x, t)}{\partial t} = -\frac{\hbar^2}{2m}\frac{\partial^2 \Psi(x, t)}{\partial x^2} + V(x)\Psi(x, t) = \hat{H}\Psi(x, t) \qquad (3.109) $$

The previous equation is called time-dependent Schrodinger's equation in terms of the wave function. Note that $V(x)$ is no longer a Hermitian operator but rather a real function.

Reconsidering a spin-1/2 system, what is the Schrodinger equation for the spin-1/2 system? Since the main interest is in understanding how the spin state of the electron changes with time due to its interaction with the static and uniform magnetic field along z-axis, therefore, for studying such a system, the focus will be on solving the Schrodinger equation for the state ket. The interaction potential between the electron and the magnetic field can be written in operator form as the following:

$$ \hat{V} = \hat{\mu}.\hat{B} \qquad (3.110) $$

Here, $\hat{V} =$ a Hermitian operator, $\hat{\mu} =$ the magnetic moment operator, and $\hat{B} =$ the magnetic field along the z-axis.

The magnetic moment of the electrons proportional to its spin, S, and is described as:

$$ \hat{\mu} = -\left(\frac{e}{m_e c}\right)\hat{S}, \qquad (3.111) $$

$$ \hat{B} = B_z k \qquad (3.112) $$

$$ \hat{V} = -\left(\frac{e}{m_e c}\right)\hat{S}. \hat{B} \qquad (3.113) $$

Since we are mainly interested in the interaction of the electron with the magnetic field along z-axis, only an interaction Hamiltonian will be considered. The Hamiltonian for such a system is simply given as:

$$\hat{H} = \hat{V} = -\left(\frac{eB_z}{m_e c}\right)\hat{S}_z \tag{3.114}$$

Before we solve the Schrodinger equation, it is helpful to determine what can be obtained by employing the Hamiltonian operator on eigenkets of spin-up and spin-down states:

$$\hat{H}|\uparrow\rangle = -\frac{\hbar}{2}\left(\frac{eB_z}{m_e c}\right)\bigg|\uparrow\rangle = -\frac{\hbar\omega}{2} \tag{3.115}$$

$$\hat{H}|\downarrow\rangle = \frac{\hbar}{2}\left(\frac{eB_z}{m_e c}\right)|\downarrow\rangle = \frac{\hbar\omega}{2} \tag{3.116}$$

where ω is known as Larmor frequency and is written as:

$$\omega = \frac{eB_z}{m_e c} \tag{3.117}$$

Since the Hamiltonian operator is dependent on the spin operator, the eigenkets of the Hamiltonian are the same as the eigenkets of the spin operator. Spin eigenkets could be used to gain information about the energy carried by the electron when it is in spin-up and spin-down states while interacting with a static magnetic field.

How do we solve the previous Schrodinger equation? There are many approaches for solving the Schrodinger equation. In the section that follows, two approaches will be discussed for solving the Schrodinger equation, namely, *time evolution operator* and *separation of variables* approaches.

3.5.2 TIME EVOLUTION OPERATOR

The state of a system at time "t" is related to its initial state by a time-evolution operator defined as:

$$|\Psi, t\rangle = U(t, 0)|\Psi, 0\rangle \tag{3.118}$$

$U(t, 0)$ is called time-evolution operator that connects state vectors at time "t" and $t = 0$.
What are the properties of this time operator?
According to the probability law, the state of a system must be normalized at all times.
Thus,

$$\langle\Psi(t)|\Psi(t)\rangle = U^*(t, 0)U(t, 0)\langle\Psi(0)|\Psi(0)\rangle = 1 \tag{3.119}$$

Therefore,

$$U^*(t, 0)U(t, 0) = 1 \tag{3.120}$$

This demonstrates that the time-evolution operator is a unitary operator. The product of such an operator with its complex conjugate must give unity due to probability conservation. Other properties of the unitary operator will be discussed in later chapters. For time being, a unitary operator will be applied to solve the Schrodinger equation.

In terms of the unitary operator, the Schrodinger equation for the state vector can be written as:

$$i\hbar \frac{\partial U(t, 0)}{\partial t} = \hat{H}U(t, 0) \tag{3.121}$$

By simply integrating the previous equation, the following unitary operator is obtained:

$$U(t, 0) = \exp\left(-\frac{i\hat{H}t}{\hbar}\right) \tag{3.122}$$

Thus,

$$|\Psi, t\rangle = \exp\left(-\frac{i\hat{H}t}{\hbar}\right)|\Psi, 0\rangle \tag{3.123}$$

Since

$$|\Psi, 0\rangle = c_\uparrow|\uparrow\rangle_z + c_\downarrow|\downarrow\rangle_z = c_\uparrow\,(t=0)|\uparrow\rangle + c_\downarrow(t=0)|\downarrow\rangle \tag{3.124}$$

and the Hamiltonian operator is independent of time, the time-dependent state vector can be written as the following:

$$|\Psi, t\rangle = c_\uparrow(t)|\uparrow\rangle + c_\downarrow(t)|\downarrow\rangle \tag{3.125}$$

The linear superposition of states is now time dependent, and any information about the state vector can be obtained for all future times. The expansion coefficients $c_\uparrow$ and $c_\downarrow$ have become time dependent and can be described as:

$$c_\uparrow(t) = c_\uparrow(0)e^{\frac{-i\omega t}{2}} \tag{3.126}$$

$$c_\downarrow(t) = c_\downarrow(0)e^{\frac{i\omega t}{2}} \tag{3.127}$$

Equation (3.125) can be rewritten as:

$$|\Psi, t\rangle = c_\uparrow(0)e^{\frac{-i\omega t}{2}}|\uparrow\rangle + c_\downarrow(0)e^{\frac{i\omega t}{2}}|\downarrow\rangle \tag{3.128}$$

This method works well, as long as time evolution of the quantum states of the system are of interest, and the Hamiltonian of the system is independent of time. To obtain the wave function from such states, all we need to do is to map the quantum state to a wave function and eigenstates to eigenfunctions. However, if we are interested in the wave function of the system, the approach that is discussed next is most frequently used.

Conceptual Question 5: *Expand the exponential in Equation (3.123) and derive the state vector in Equation (3.128).*

3.5.3 SEPARATION OF VARIABLES: TIME-INDEPENDENT SCHRODINGER EQUATION

The time-dependent Schrodinger equation for the wave function is:

$$i\hbar\frac{\partial\Psi(x, t)}{\partial t} = -\frac{\hbar^2}{2m}\frac{\partial^2\Psi(x, t)}{\partial x^2} + V(x)\Psi(x, t) \tag{3.129}$$

In the previous equation, $V(x)$ is independent of time. For this case, the Schrodinger equation can be solved by the method of separation of variables. In this method, the wave function can be simply expressed as a product of the spatial and temporal functions shown as:

$$\Psi(x, t) = \psi(x)\varphi(t) \tag{3.130}$$

$\psi(x)$ is a function of spatial coordinate "x" alone, and $\varphi(t)$ is a function of time "t" alone. With these functions, the Schrodinger equation takes the following form:

$$i\hbar\,\psi(x)\frac{d\varphi(t)}{dt} = -\frac{\hbar^2}{2m}\varphi(t)\frac{d^2\psi(x)}{dx^2} + V(x)\psi(x)\varphi(t) \tag{3.131}$$

Dividing the previous equation by $\psi(x)\varphi(t)$

$$i\hbar\,\frac{1}{\varphi}\frac{d\varphi(t)}{dt} = -\frac{\hbar^2}{2m}\frac{1}{\psi}\frac{d^2\psi(x)}{dx^2} + V(x) \tag{3.132}$$

The left side of the equation is a function of "t" alone, and the right side is a function of "x" alone and are independent of each other. They must be equal to some constant value. Both sides of the equations can be equated to some constant value E:

$$i\hbar\,\frac{1}{\varphi}\frac{d\varphi(t)}{dt} = E \tag{3.133}$$

$$i\hbar\,\frac{d\varphi(t)}{dt} = E\varphi \tag{3.134}$$

and

$$-\frac{\hbar^2}{2m}\frac{1}{\psi}\frac{d^2\psi(x)}{dx^2} + V(x) = E \tag{3.135}$$

We get:

$$-\frac{\hbar^2}{2m}\frac{d^2\psi(x)}{dx^2} + V(x)\psi = E\psi \tag{3.136}$$

The previous equation is called a *time-independent Schrodinger equation* and is a second-order ordinary differential equation. For many systems where the potential function is independent of time, separation of variables method can be applied to solve the Schrodinger equation. The previous equation can be solved by imposing boundary conditions, which will be discussed in the next chapter.

3.5.4 STATIONARY STATES

The solution of the time-dependent part of the Schrodinger Equation (3.134) is simple and written as:

$$\varphi(t) = e^{-iEt/\hbar} \tag{3.137}$$

From Equation (3.130), the wave function can now be written as:

$$\Psi(x, t) = \psi(x)e^{-iEt/\hbar} \tag{3.138}$$

and the probability density of the previous wave function is:

$$\rho(x, t) = |\Psi(x, t)|^2 = \psi(x)\psi(x)^* = \text{constant} \tag{3.139}$$

Thus, the probability density of the wave function is independent of time. Such quantum states are called *stationary states*. As the word stationary implies, the probability of finding the particle at all points in space will remain constant in time, as if the particle is stationary, or in other words, "the particle is frozen in time." In these states, the measurement of a particle's energy would yield an exact value. Such states are also called energy eigenstates and determinate states.

EXAMPLE 3.4

Determine the probability densities of the following wave functions and determine whether they are stationary states.

$$\Psi(x, t) = Ae^{\frac{-i\omega t}{2}}\psi_1(x) \tag{3.140}$$

$$\Psi(x, t) = Ae^{\frac{-i\omega t}{2}}\psi_1(x) + Be^{\frac{i\omega t}{2}}\psi_2(x) \tag{3.141}$$

where A and B are some real numbers and constants.

$$\rho(x, t) = |\Psi(x, t)|^2 = A^2\psi_1^2 = \text{constant} \tag{3.142}$$

$$\rho(x, t) = |\Psi(x, t)|^2 = A^2\psi_1^2 + B^2\psi_2^2 + AB\,(e^{-i\omega t} + e^{i\omega t}) \tag{3.143}$$

$$\rho(x, t) = |\Psi(x, t)|^2 = A^2\psi_1^2 + B^2\psi_2^2 + 2\,AB\,\cos\,(\omega t) \neq \text{constant} \tag{3.144}$$

The probability density of the wave function in Equation (3.142) is independent of time. Thus, it is a stationary state, whereas the probability density of the wave function Equation (3.144) is dependent on time. Thus, it is not a stationary state.

PROBLEMS

3.10 The wave function of a two-state system at time $t = 0$ is given by $\Psi(x, 0) = c_0 u_0(x) + c_1 u_1(x)$. If the energies of the system in two states are E_0 and E_1, find its wave function at a later time "t" by applying a unitary operator. You can assume the interaction Hamiltonian to be time independent. Also, find the probability density at a later time "t." Finally, compare the probability density at time $t = 0$ and at time "t." Do you see any difference? If yes, describe it.

3.11 The Gaussian wave function of a particle at time $t = 0$ is given by, $\Psi(x, 0) = Ae^{-\gamma x^2}$ where, A and γ are positive constants. Find the expectation value of $\hat{H}$ at time $t = 0$.

3.12 Two approaches (time-evolution operator and separation of variables) are discussed in this chapter for solving the Schrodinger equation. Describe in your own words what information about the quantum system is extracted by applying these two methods?

3.13 What is common between a stationary state, eigenstate and determinate state? How does the probability densities of these states change with time?

3.6   CONSERVATION OF PROBABILITY

In classical mechanics, physical variables such as momentum and energy of a system are conserved quantities when the system is closed and isolated, which means that these variables remain constant for all times. In quantum mechanics, one important quantity is the probability of locating the particle in some region of space. Thus, we must find out whether probability of a system is a conserved quantity.

The probability of a system is:

$$P = \int \psi^*(x, t)\psi(x, t)\, dx \tag{3.145}$$

For probability to be conserved, it is required that:

$$\frac{\partial P}{\partial t} = \int \frac{\partial(\psi^*(x, t)\psi(x, t))}{\partial t}\, dx = 0 \tag{3.146}$$

Let us see whether we can prove it.

$$\int \frac{\partial(\psi^*\psi)}{\partial t}\, dx = \int \frac{\partial \psi^*}{\partial t}\psi\, dx + \int \frac{\partial \psi}{\partial t}\psi^*\, dx \tag{3.147}$$

We know that the Schrodinger equation is:

$$i\hbar\, \frac{\partial \psi(x, t)}{\partial t} = H\, \psi(x, t) \tag{3.148}$$

where H is the Hamiltonian operator of the system.

The complex conjugate of it is:

$$-\, i\hbar\, \frac{\partial \psi^*(x, t)}{\partial t} = H^*\psi^*(x, t) \tag{3.149}$$

Thus, by substituting Equations (3.148) and (3.149) into Equation (3.147), we get:

$$\int \frac{\partial(\psi^*\psi)}{\partial t}\, dx = \frac{i}{\hbar} \int (\psi H^*\psi^* - \psi^* H\, \psi)\, dx \tag{3.150}$$

For the previous integral to be equal to zero, it is required that:

$$H^*\psi^* = E\psi^*; \quad H\psi = E\psi \tag{3.151}$$

where 'E' is some real number ($E = E^*$), which means that H must be a Hermitian operator ($H^* = H$). By substituting Equation (3.151) into the Equation (3.150), the following is obtained:

$$\frac{\partial P}{\partial t} = \int \frac{\partial(\psi^*\psi)}{\partial t}\, dx = \frac{i}{\hbar} \int E(\psi\psi^* - \psi^*\psi)\, dx = 0 \tag{3.152}$$

$$P = \text{const.} \tag{3.153}$$

Thus, for the probability of a quantum system to be conserved, it is required that the Hamiltonian of a system must be a Hermitian operator.

PROBLEM

3.14 For the states given,

$$\Psi(x, t) = Ae^{\frac{-i\omega t}{2}}\psi_1(x) \tag{3.140}$$

$$\Psi(x, t) = Ae^{\frac{-i\omega t}{2}}\psi_1(x) + Be^{\frac{i\omega t}{2}}\psi_2(x) \tag{3.141}$$

Show that probability of a quantum system corresponding to these stationary and nonstationary states is a conserved quantity.

3.7 HEISENBERG UNCERTAINTY PRINCIPLE

We began this chapter by showing that "indeterminacy" associated with quantum mechanical systems is what makes quantum mechanics so different from classical mechanics. The fundamental principle that describes this indeterminacy is called Heisenberg's uncertainty principle. This principle describes the soul of quantum mechanics. Without this principle, quantum mechanics would be determinate and just classical mechanics. It tells us that all physical quantities that can be observed are exposed to unpredictable fluctuations, so that their values cannot be precisely determined simultaneously. This uncertainty is inherent in nature and is not due to lack of information about the system. Some other fields such as the stock market, a weather forecast and economics also have uncertainty in their systems, but such uncertainty is mainly due to lack of information. This indeterminacy associated with quantum mechanical systems cannot be overcome by building technologically advanced devices. Therefore, quantum mechanics is a statistical theory and can make definite predictions of an ensemble of identical systems but nothing definite about an individual system.

In the section that follows, the mathematical foundation of the Heisenberg's uncertainty principle is discussed.

Consider two physical observables "M" and "P" of a system. These physical observables are represented as operators, $\hat{M}$ and $\hat{P}$. The measurement of these observables will have uncertainty in their values. This uncertainty in their values can be calculated using operator of standard deviation, and is given in the following form as:

$$\Delta\hat{M} = \hat{M} - \langle\hat{M}\rangle \tag{3.154}$$

where the expectation value is taken over a physical state under consideration. The expectation value of the square of the operator of standard deviation is called as uncertainty in the value of observable "M" and is:

$$\langle(\Delta\hat{M})^2\rangle = \langle(\hat{M}^2 - 2\hat{M}\langle\hat{M}\rangle + \langle\hat{M}\rangle^2)\rangle = \langle\hat{M}^2\rangle - \langle\hat{M}\rangle^2 \tag{3.155}$$

Similarly, for P:

$$\langle(\Delta P)^2\rangle = \langle\hat{P}^2\rangle - \langle\hat{P}\rangle^2 \tag{3.156}$$

According to the Heisenberg's uncertainty principle:

$$\langle(\Delta\hat{M})^2\rangle\langle(\Delta P)^2\rangle \geq \left(\frac{1}{2i}\langle[\hat{M}, \hat{P}]\rangle\right)^2 \tag{3.157}$$

where $[\hat{M}, \hat{P}] = \hat{M}\hat{P} - \hat{P}\hat{M}$ is the commutator bracket. These uncertainties depend on commutation of observables. If the commutator bracket of any two observables is zero, then there is no uncertainty in the measurement of their simultaneous values. Such observables have simultaneous eigenfunctions and are called compatible observables. The observables whose commutator is non-zero are called incompatible observables. For example, position and momentum are incompatible observables, and using uncertainty principle, the position and momentum uncertainties are:

$$\langle(\Delta\hat{x})^2\rangle\langle(\Delta p)^2\rangle \geq \left(\frac{1}{2i}\langle[\hat{x},\hat{p}]\rangle\right)^2 \tag{3.158}$$

$$[\hat{x},\hat{p}] = i\hbar \tag{3.159}$$

Thus,

$$\langle(\Delta\hat{x})^2\rangle\langle(\Delta p)^2\rangle \geq \frac{\hbar^2}{4} \tag{3.160}$$

If we denote the standard deviation or uncertainty in position and momentum as:

$$\sigma_x^2 = \langle(\Delta\hat{x})^2\rangle \text{ and } \sigma_p^2 = \langle(\Delta\hat{p})^2\rangle,$$

then the uncertainty principle using these notations can be written as:

$$\sigma_x^2\sigma_p^2 \geq \frac{\hbar^2}{4} \tag{3.161}$$

In some of the books, the following notation is used for standard deviation, $\sigma_x = \Delta x$ and $\sigma_p = \Delta p$, and using this notation:

$$\Delta x\,\Delta p \geq \frac{\hbar}{2} \tag{3.162}$$

The previous equation means that if a measurement of position is made with accuracy Δx, and measurement of momentum is made simultaneously with accuracy of Δp, the product of two will never be smaller than ($\hbar/2$).

Similarly, the energy and time uncertainty relationship is given as:

$$\Delta E\Delta t \geq \frac{\hbar}{2} \tag{3.163}$$

Since time "t" is not an observable rather an independent variable of which dynamical quantities are functions, therefore, Δt is not the error in the measurement of time, but it is the time it takes for a system to change significantly. To understand the energy-time uncertainty relationship, consider an atom that absorbs a quantum of energy from a radiation field during a time-interval Δt. The energy transferred to the atom is ΔE, and transfer of this energy takes place within a time interval Δt. Thus, the product of energy transferred and the time interval during which energy is transferred is always going to be greater than a smaller number given in Equation (3.163). If we know precisely how much energy is transferred (ΔE) to the atom, then we are uncertain about the time taken (Δt) for the transfer of energy.

Equation (3.162) can be generalized to other components of momentum and position.

$$\Delta y \, \Delta p_y \geq \frac{\hbar}{2} \tag{3.164}$$

$$\Delta z \, \Delta p_z \geq \frac{\hbar}{2} \tag{3.165}$$

3.7.1 PROOF OF THE UNCERTAINTY PRINCIPLE

3.7.1.1 The Schwarz Inequality

In Euclidean space, any two vectors "a" and "b" obey the following inequality called as the Schwarz inequality:

$$|a|^2 |b|^2 \geq |a. \, b|^2 \tag{3.166}$$

Similarly, in a complex vector space, for two vectors, $|\psi\rangle$, $|\phi\rangle$, the Schwarz inequality can be written as:

$$\langle \psi | \psi \rangle \langle \phi | \phi \rangle \geq \left| \langle \psi | \phi \rangle \right|^2 \tag{3.167}$$

Let us now consider two observables $\hat{A}$ and $\hat{B}$, and their operators of standard deviation (uncertainty) are $\Delta\hat{A}$ & $\Delta\hat{B}$. Using Schwarz's inequality, the product of these operators can be written as:

$$\langle (\Delta\hat{A})^2 \rangle \langle (\Delta\hat{B})^2 \rangle \geq |\langle \Delta\hat{A}.\Delta\hat{B} \rangle|^2 \tag{3.168}$$

The commutator for observables $\hat{A}$ and $\hat{B}$ can be written as:

$$[\hat{A}, \hat{B}] = \langle \psi | \hat{A}\hat{B} | \psi \rangle - \langle \psi | \hat{B}\hat{A} | \psi \rangle$$
$$= \langle \hat{A}\hat{B} \rangle - \langle \hat{B}\hat{A} \rangle \tag{3.169}$$

Also, the anti-commutator of $\hat{A}$ and $\hat{B}$ can be written as:

$$\{\hat{A}, \hat{B}\} = \langle \psi | \hat{A}\hat{B} | \psi \rangle + \langle \psi | \hat{B}\hat{A} | \psi \rangle$$
$$= \langle \hat{A}\hat{B} \rangle + \langle \hat{B}\hat{A} \rangle \tag{3.170}$$

Thus, by adding the previous two equations, we get:

$$\langle \hat{A}\hat{B} \rangle = \frac{[\hat{A}, \hat{B}]}{2} + \frac{\{\hat{A}, \hat{B}\}}{2}. \tag{3.171}$$

Similarly, for standard deviation operators of the previous operators, the product can be written as:

$$\langle \Delta\hat{A}.\Delta\hat{B} \rangle = \frac{[\Delta\hat{A}, \Delta\hat{B}]}{2} + \frac{\{\Delta\hat{A}, \Delta\hat{B}\}}{2} \tag{3.172}$$

By getting rid of anti-commutator, we can have:

$$\langle \Delta\hat{A}.\Delta\hat{B} \rangle \geq \frac{[\Delta\hat{A}, \Delta\hat{B}]}{2} \tag{3.173}$$

But,

$$[\Delta\hat{A}, \Delta\hat{B}] = [\hat{A}, \hat{B}] \tag{3.174}$$

Therefore, using Equation (3.174), we can state the Schwarz inequality Equation (3.168) for the previous standard deviation operators as:

$$\langle(\Delta\hat{A})^2\rangle\langle(\Delta\hat{B})^2\rangle \geq \left|\frac{[\hat{A}, \hat{B}]}{2}\right|^2 \tag{3.175}$$

or

$$\langle(\Delta\hat{A})^2\rangle\langle(\Delta\hat{B})^2\rangle \geq \left|\frac{[\hat{A}, \hat{B}]}{2i}\right|^2 \tag{3.176}$$

This proof shows that the uncertainty principle can be simply obtained from the mathematical structure of quantum mechanics.

> The uncertainty principle refers to the degree of indeterminateness in the possible knowledge of the simultaneous values of various quantities with which the quantum theory deals; it does not restrict, for example, the exactness of a position measurement alone or a velocity measurement alone.
>
> **Werner Heisenberg**
> *The Physical Principles of the Quantum Theory*

3.7.2 Applications of the Principle

3.7.2.1 Heisenberg's Microscope

The resolving power of the microscope, which is simply the "ability of the microscope to yield distinct images of points that are situated very close to one another," is known to provide an accuracy:

$$\Delta x = \frac{\lambda}{\sin(\theta)} \tag{3.177}$$

in position determination, where λ is the wavelength of the radiation that enters the lens, and θ is the half angle subtended at the particle at position P, as shown in the Figure 3.6.

Let us suppose that a microscope is used to locate a particle at position P by using light. Figure 3.6 clearly illustrates such a microscope. When looking through such a microscope at a particle located at P, we cannot conclude that this particle is precisely at P, because if it were slightly situated to the left or to the right of point P, its image would still appear the same. The image appears to be the same because of limits imposed on the resolving power of the microscope, as shown in Equation (3.177). Hence, in trying to locate a particle to point P, we will always be confronted with uncertainty in its location given by Equation (3.177). We will observe point P as a dot whose diameter is Δx.

Let the particle be moving along x-axis, having momentum $p_x = mv_x$ before it is in the field of vision of the microscope. After we have seen such a particle as a dot in the microscope, its momentum has changed due to its interaction with the light. This interaction is simply the collision between the photon of light and the particle. Such a collision may be likened to a collision between two perfectly elastic balls, the photon getting a kick from the particle and is deflected. This kick is called Compton recoil and cannot be exactly known, since the direction of the scattered photon is undetermined within the bundle of rays entering the microscope.

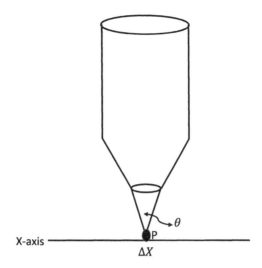

FIGURE 3.6 This schematic illustrates Heisenberg's microscope.

The changed momentum of the particle along x-axis is:

$$\Delta p_x = \frac{h}{\lambda} \sin(\theta) \qquad (3.178)$$

The changed momentum is the uncertainty in the momentum of the particle after the observation. Thus, if we now multiply the previous two uncertainties, we get:

$$\Delta x \, \Delta p_x = h \qquad (3.179)$$

This result will always be the same regardless of the microscope used and regardless of the wavelength used for illuminating the particle. This experiment clearly shows that any attempt to determine the position of the particle will cause disturbance to its momentum. Therefore, momentum and position of the particle cannot be simultaneously determined. This imprecision in the measurement of position and momentum is not due to lack technologically advanced devices but rather due to theoretical foundation of quantum mechanics. Heisenberg's uncertainties are the theoretical limits to the measurement.

3.7.2.2 Defining Orbits in Atoms

Any attempt to determine the path of an electron revolving around the nucleus of an atom run into difficulties. This is mainly due to the uncertainty principle. To determine the path of an electron, one needs to find the location of the electron at a particular instance of time, but any attempt to locate the electron accurately will disturb its momentum largely. As was discussed in the previous section on the microscope, to locate an electron, radiation must be shined on such an electron. The interaction between the radiation and the electron will disturb its momentum in such a way that the electron will be knocked out of its path and thrown into the next orbit. Thus, it is impossible to follow a particle as it moves in an orbit by watching it with a microscope because the quanta of radiation used in observing it will not only send the electron into some other orbit but into one that cannot be predicted and controlled. In other words, according to the uncertainty principle, an electron moving around the nucleus does not have a well-defined path as opposed to a classical view.

PROBLEMS

3.15 In Equation (3.168), the Schwarz inequality for the standard deviation operators $\left(\Delta \hat{A}\right)^2$ & $\left(\Delta \hat{B}\right)^2$ is discussed. If we replace these operators with $\Delta \hat{A}$ & $\Delta \hat{B}$, what would be the Schwarz inequality? Find the expectation values of operators $\Delta \hat{A}$ & $\Delta \hat{B}$.

3.16 Prove that $\left[\Delta \hat{A}, \Delta \hat{B}\right] = \left[\hat{A}, \hat{B}\right]$.

3.17 Explain the physical meaning of position-momentum and energy-time uncertainty relationships using thought experiments.

3.18 Evaluate $\left[\hat{x}^2, \hat{p}^2\right]$, and derive the Schwarz inequality for these operators.

3.19 What is the main difference between the unpredictability of quantum mechanics and the stock market? Explain.

3.20 Fill out the flow chart with the appropriate boxes given in Figures 3.7 and 3.8.

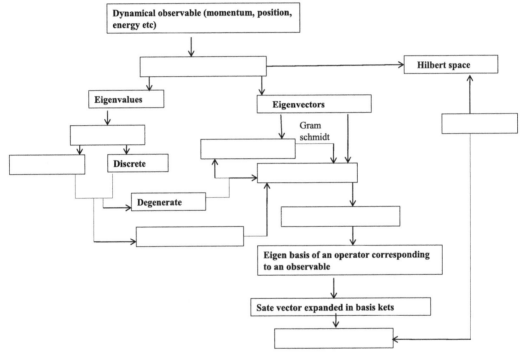

FIGURE 3.7

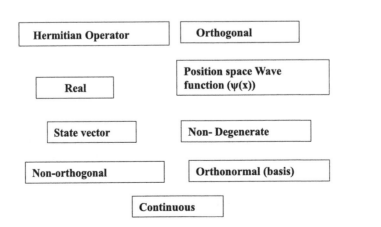

FIGURE 3.8

3.8 MATRIX MECHANICS

The two equivalent mathematical theories that laid the foundation of the theory of quantum mechanics are wave mechanics and matrix mechanics. Wave mechanics is rooted in classical ideas of particle and wave and, thus, seems easier to grasp. At the same time, the complexity of the quantum phenomenon challenges such ideas. For example, the wave function of a quantum particle represents a probability wave associated with the particle, and the absolute square of the wave function gives the probability density of the particle and is of great significance. However, the absolute of the wave function, a probability amplitude, is not of much significance. Experimentally, directly one can only determine probability density and not probability amplitude. If probability amplitude cannot be determined experimentally, then its meaning becomes absurd. However, for waves such as an elastic wave or sound waves, the absolute amplitude determines the intensity of the note and is of great significance. This contrast between probability amplitude of the wave associated with the quantum particle and the absolute amplitude of the sound wave may give rise to confusion. Matrix mechanics, on the other hand, does not provide any way of direct picturization of the physical phenomenon. It is more abstract, mathematically cumbersome, and eliminates the possibility of any misconception. Yet, for some problems, wave mechanics provides easier solutions, while for others, matrix mechanics may be a better approach. In this book, we have purposely used wave mechanics throughout the entire book to avoid confusion arising out of usage to the two methods simultaneously. Here, we give a brief introduction to matrix mechanics to help the students gain familiarity with it.

What is a matrix?

A matrix is a rectangular array of numbers or symbols, arranged in row and columns. The size of a matrix is defined by the number of rows and columns it contains. A matrix with n rows and m columns is called $n \times m$ matrix, and rows and columns are called dimensions. Thus, $n \times m$ matrix has $n \times m$ dimensions. For example,

$$\begin{bmatrix} 0 & 1 \\ 1 & 0 \end{bmatrix} \tag{3.180}$$

is a 2×2 matrix. Each number in the matrix is called matrix element.

Matrices that have a single row are called row matrices, and matrices that have a single column are column matrices.

A vector in a vector space can be represented as a matrix. Let us now go back to the superposition state of the spin state of the electron that we discussed at the beginning of this chapter.

$$|\Psi\rangle = c_\uparrow |\uparrow\rangle_z + c_\downarrow |\downarrow\rangle_z \tag{3.5}$$

How do we represent such a state vector as matrix?

We know that spin-up and spin-down vectors lie in a 2D vector space. Thus, spin-up ket vector can be represented as:

$$|\uparrow\rangle_z = \begin{bmatrix} 1 \\ 0 \end{bmatrix} \tag{3.181}$$

which is a 2D column matrix and is also called as spinor (fancy name!).

Similarly,

$$|\downarrow\rangle_z = \begin{bmatrix} 0 \\ 1 \end{bmatrix} \tag{3.182}$$

The state vector can now be written as:

$$|\Psi\rangle = c_\uparrow \begin{bmatrix} 1 \\ 0 \end{bmatrix} + c_\downarrow \begin{bmatrix} 0 \\ 1 \end{bmatrix} \tag{3.183}$$

$$\text{Thus,} \quad |\Psi\rangle = \begin{bmatrix} c_\uparrow \\ c_\downarrow \end{bmatrix} \tag{3.184}$$

You can check here that $|\downarrow\rangle_z$ & $|\uparrow\rangle_z$ remains orthogonal. Try doing the matrix multiplication of the previous two column matrices. For $|\Psi\rangle$ as a state vector in a vector space, each $c_\uparrow$ & $c_\downarrow$ represent components of such a vector.

Not only vectors but operators can also be represented as matrices. For example, $\hat{S}_z$ can be represented as a 2×2 matrix,

$$\hat{S}_z = \frac{\hbar}{2} \begin{bmatrix} 1 & 0 \\ 0 & -1 \end{bmatrix} \tag{3.185}$$

$$\hat{S}_z|\uparrow\rangle_z = \frac{\hbar}{2} \begin{bmatrix} 1 & 0 \\ 0 & -1 \end{bmatrix} \begin{bmatrix} 1 \\ 0 \end{bmatrix} = \frac{\hbar}{2} \begin{bmatrix} 1 \\ 0 \end{bmatrix} = \frac{\hbar}{2}|\uparrow\rangle_z \tag{3.186}$$

Here $+(\hbar/2)$ is the eigenvalue for the $\hat{S}_z$ corresponding to the spin-down ket vector, also called as eigenvector of operator $\hat{S}_z$. Try doing the same steps for the spin-down ket vector.

Thus, in matrix mechanics, both state vectors and operators are represented as matrices. Representing multidimensional systems using this approach becomes easier.

The X and Y components of the spin vector can be represented as the following matrices:

$$\hat{S}_x = \frac{\hbar}{2} \begin{bmatrix} 0 & 1 \\ 1 & 0 \end{bmatrix}, \quad \hat{S}_y = \frac{\hbar}{2} \begin{bmatrix} 0 & -i \\ i & 0 \end{bmatrix} \tag{3.187}$$

The eigenvectors of operators, $\hat{S}_x$ and $\hat{S}_y$ are given as:

$$|\uparrow\rangle_x = \frac{1}{\sqrt{2}} \begin{bmatrix} 1 \\ 1 \end{bmatrix}, \quad |\downarrow\rangle_x = \frac{1}{\sqrt{2}} \begin{bmatrix} 1 \\ -1 \end{bmatrix} \tag{3.188}$$

$$|\uparrow\rangle_y = \frac{1}{\sqrt{2}} \begin{bmatrix} 1 \\ i \end{bmatrix}, \quad |\downarrow\rangle_y = \frac{1}{\sqrt{2}} \begin{bmatrix} 1 \\ -i \end{bmatrix} \tag{3.189}$$

We will not go into further development of the concepts of matrix mechanics and end our discussion here.

3.9 TUTORIALS

3.9.1 SPIN-1/2

Purpose

To understand electron's spin, and how it changes with time in the presence of a uniform magnetic field using Bloch vector.

Concepts

- The intrinsic angular momentum of a particle is called spin. Such an observable is denoted as $\hat{S}$.
- The experimentally measured values of an electron's spin are $+\frac{\hbar}{2}$ (spin-up) and $-\frac{\hbar}{2}$ (spin-down) where $\hbar$ is Planck's constant.
- An electron has magnetic moment due to its spin, and is given as $\hat{\mu} = -\left(\frac{e}{m_e c}\right)\hat{S}$.
- The interaction between the electron's magnetic moment and external uniform static magnetic field causes the electron's spin precession about the axis of magnetic field with a frequency called as Larmor frequency ($\omega = \frac{eB_z}{m_e c}$).
- A Bloch sphere is a visualization tool for the spin state of a particle. Using Bloch sphere any superposition state of the spin of the electron can be written as,

$$|\psi\rangle = \cos\frac{\theta}{2}|\uparrow\rangle + \sin\frac{\theta}{2}e^{i\varphi}|\downarrow\rangle.$$

Question 1 In an experiment, a beam of N electrons is made to pass through a uniform magnetic field and two different values of the spin are measured. Write a mathematical expression for the state of the electrons before the measurement. In the measurements, half of the electrons yield $+\frac{\hbar}{2}$ *(spin-up)* value and remaining half yield $-\frac{\hbar}{2}$ *(spin-down)*. Describe the states of the electrons after the measurement.

_____

_____

_____

_____

_____

Question 2 In an experiment, a beam of N electrons is made to pass through a uniform magnetic field and five different values of the spin are measured. Write a mathematical expression for the state of the electrons before the measurement.

_____

_____

_____

_____

_____

Question 3   In an experiment only one value of the spin is measured. Write a mathematical expression for the state of the electrons before the measurement.

_____

_____

_____

_____

_____

Question 4   Write the Schrodinger's equation for an electron interacting with a uniform magnetic field. Determine its solution.

_____

_____

_____

_____

_____

Use the following simulation to answer the following questions, https://www.st-andrews.ac.uk/physics/quvis/simulations_html5/sims/blochsphere/blochsphere.html

Question 5   The time-dependent state of the spin of the electron can be written as,

$$|\Psi, t\rangle = c_\uparrow(0)e^{-\frac{i\omega t}{2}}|\uparrow\rangle + c_\downarrow(0)e^{\frac{i\omega t}{2}}|\downarrow\rangle$$

Show that such a state can be written equivalently as a Bloch vector state as,

$$|\psi\rangle = \cos\frac{\theta}{2}|0\rangle + \sin\frac{\theta}{2}e^{i\varphi}|1.\rangle$$

Is angle φ a constant or time dependent variable? Determine such angle in terms of Larmor frequency.

_____

_____

_____

_____

_____

Question 6 Use the following simulation, https://www.st-andrews.ac.uk/physics/quvis/simulations_html5/sims/bloch-timedev/bloch-timedev.html and study the different states shown on the left side of the window. Answer the following questions,

    a. Is the Larmor frequency same for all the states? Why or why not? Explain.
    b. Are there any stationary states? Describe.
    c. Does the electron continue to precess if there is no external magnetic field?
    d. Explain why spin eigenstates are called determinate states and superposition states of the spins eigenstates are called indeterminate states.

_____

_____

_____

_____

_____

3.9.2 WAVE FUNCTION

Purpose

To understand the difference between any classical function and wave function of a quantum mechanical system.

Concepts

- Classically, dynamical variables such as position, momentum and energy of a system are described as functions of time as, $x(t)$, $p(t)$ and $E(t)$.
- Such functions are determined by solving dynamical equations of the system using boundary conditions.
- Quantum mechanically, dynamical variables of a system are described as operators, such as $\hat{x}$, $\hat{p}$ and $\hat{H}$.
- The Schrodinger equation of a quantum mechanical system is a dynamical equation. By solving such equation using boundary conditions, a wave function of the system is determined.
- Using such wave function, the averages of the dynamical variables of a quantum mechanical system are determined.

Question 1 Consider a particle of mass "m" moving along the x-axis, the force acting on the particle is given as:

$$F = m\,a$$

The acceleration of the particle is constant, and the boundary conditions are $t = 0$, $x(0) = 0$, $v(0) = 0$. Determine the position, momentum and total energy functions of the particle. Show that all these variables are functions of time.

_____

_____

_____

_____

_____

Question 2   Consider a quantum mechanical particle of mass "m" under the influence of 1-dimensional potential $V(x)$ (force). It comprises of two discrete energy levels. The energy eigenvalues are E_1 and E_2, and ϕ_1 and ϕ_2 are the energy eigenvectors. Describe the state vector of such a system. Explain the meaning of the expansion coefficients.

_____

_____

_____

_____

_____

What are its basis vectors? Write the completeness relation for such basis vectors.

_____

_____

_____

_____

_____

Question 3   Write the mathematical steps leading to the wave function for such a system from the previous equation of the state vector. Describe the energy eigenfunctions.

_____

_____

_____

_____

_____

Question 4 Explain the meaning of the wave function. What is the use of the wave function for a quantum mechanical system?

_____

_____

_____

_____

_____

Question 5 Write the time-dependent Schrodinger equation for such a system using wave function.

_____

_____

_____

_____

_____

Question 6 What method you would use to solve the previous equation? Describe such a method.

_____

_____

_____

_____

Question 7 Write the time-independent Schrodinger equation. Describe the time-independent wave function.

_____

_____

_____

_____

_____

Question 8 What would you do to obtain information about the dynamical variables such as position, momentum and energy of the quantum particle?

_____

_____

_____

_____

_____

Question 9 What do you understand by the expectation value of a dynamical variable such as the position of a quantum mechanical system? Write the mathematical expression for the expectation value of the position variable of the previous quantum mechanical system using the wave function.

_____

_____

_____

_____

_____

Question 10 Write the wave function of the previous quantum mechanical system at some later time "t" if it started initially in one of the energy eigenstates.

_____

_____

_____

_____

_____

Question 11 Write the wave function of the previous quantum mechanical system at some later time "t" if it started initially in a linear superposition state.

_____

_____

_____

_____

_____

Question 12 Consider the following wave function, $\Psi(x) = c_1\varphi_1(x) + c_2\varphi_2(x)$ of the particle. Determine the expectation value of the energy of the particle.

_____

_____

_____

_____

_____

Question 13 Consider the following wave function, $\Psi(x) = \varphi_1(x)$ of the particle. Determine the expectation value of the energy of the particle. Is there any difference between the expectation values of the energy of the particle when compared with the expectation value of the energy in Question 12?

_____

_____

_____

_____

_____

Question 14 Describe in your own word the differences between the quantum mechanical and classical approaches.

_____

_____

_____

_____

_____

4 Applications of the Formalism-I

4.1 INTRODUCTION

In this chapter, several applications of the formalism of theory of quantum mechanics are discussed. The time-independent Schrodinger equation is applied to understand simple quantum systems such as a free particle, a particle in an infinite square well, step potentials, potential barrier and finite square well.

4.2 THE FREE PARTICLE

In this section, we begin with the simplest case, a free particle. It is the most elementary application of the formalism. A free particle means that no forces are acting on the particle, and hence, there is no interaction between the particle and its environment.

For simplicity, we will consider only a 1-dimensional system. Consider a particle of mass "m" moving along x-axis. Since no forces are acting on the particle, the potential energy function of the particle is zero as shown:

$$V(x) = 0 \tag{4.1}$$

The Hamiltonian operator for the particle is

$$\hat{H} = -\frac{\hbar^2}{2m}\frac{d^2}{dx^2} \tag{4.2}$$

The total energy of the particle is its kinetic energy and is positive. Since there are no bounds to restrict the energy of the particle, it can have any positive values. These energy values are not discrete, but rather they are continuous.

What aspects are of interest in the study of the particle?

The first issue to consider is the motion of the particle and to determine the properties of the motion.

According to classical mechanics, a free particle moves with a constant velocity along a straight line. Its kinetic energy and momentum are constant in time. A key question is whether or not a free quantum particle behaves in a similar way.

To determine the motion of the quantum particle, the Schrodinger equation must be solved for the free particle to obtain the wave function. By studying how the probability density of the particle changes with time, it will be possible to understand its motion. Using the wave function, the average energy and the momentum of the particle can be calculated.

The wave function of the particle is described as:

$$\langle x|\Psi\rangle = \Psi(x) \tag{4.3}$$

Using the method of separation of variables, the Schrodinger equation of the particle can be solved for the spatial part, and then the time-dependent function can be included to obtain the time-dependent wave function of the particle.

The Schrodinger equation of the particle is:

$$\hat{H}\Psi(x) = E\Psi(x) \tag{4.4}$$

$$\frac{d^2\Psi(x)}{dx^2} = -k^2\,\Psi(x), \quad k^2 = \frac{2mE}{\hbar^2} \;\&\; k = \mp\frac{\sqrt{2mE}}{\hbar} \tag{4.5}$$

The previous equation is the same as a time-independent classical wave equation. This also explains the use of the term "wave equation" for the Schrodinger equation. The solution of the previous equation is:

$$\Psi(x) = \alpha e^{ikx} + \beta e^{-ikx} \tag{4.6}$$

where α and $\beta =$ arbitrary constants.

In the previous equation, the two eigenfunctions are αe^{ikx} and βe^{-ikx}, corresponding to the two possible directions of the wave vector "k" or momentum for a given value of energy "E" of the particle. The positive exponential term represents a plane wave moving to the right, and the negative exponential represents a plane wave moving to the left. Thus, the wave function of the particle is a superposition of waves moving to the right and left.

The time-dependent wave function can be obtained simply by adding a factor of $\exp\left(-iEt/\hbar\right)$ to the wave function.

$$\Psi(x,\,t) = \alpha e^{i(kx-\frac{E}{\hbar}t)} + \beta e^{-i(kx+\frac{E}{\hbar}t)} \tag{4.7}$$

The previous wave function represents a superposition of plane waves traveling in space-time. The interesting thing about the previous wave function is that it represents plane waves associated with the particle. This, as we discussed in Chapter 2, was predicted by Louis de Broglie, as de Broglie waves associated with a quantum particle. According to him, a quantum mechanical particle has waves associated with it as it moves in space. The momentum, energy and wavelength of such waves is defined as:

$$p = \hbar k, \quad E = \frac{\hbar^2 k^2}{2m} = \frac{p^2}{2m}, \quad \lambda = \frac{2\pi}{|k|} \tag{4.8}$$

As shown previously, the energy of the wave associated with the free quantum particle is same as that of the classical free particle.

The next step is to determine whether or not the wave function of the particle is normalizable. For simplicity, consider a particle moving to right only and its wave function as:

$$\Psi(x,\,t) = \alpha e^{i(kx-\frac{E}{\hbar}t)} \tag{4.9}$$

Since the probability density of the particle will remain constant with time, such a wave function is called a "stationary state" wave function. Normalization of the previous wave function is:

$$\int_{-\infty}^{\infty} \Psi(x,\,t)\Psi(x,\,t)^*\,dx = |\alpha|^2 \int_{-\infty}^{\infty} dx = \infty \tag{4.10}$$

Thus, the wave function is not normalizable. The wave function of the particle is not physically realizable, but this does not mean that the wave function of the particle is meaningless. To make sense of the wave function, the position space wave function is expressed in its momentum space wave function form, as is shown next. Any position-space wave function is related to its momentum-space wave function form by the following equation which is an inverse Fourier transform of the momentum space wave function to the position space wave function.

$$\Psi(x, t) = \frac{1}{\sqrt{2\pi}} \int_{-\infty}^{\infty} \emptyset(k) e^{i(kx - \frac{Et}{\hbar})} dk \tag{4.11}$$

This wave function can be normalized for a range of k's and energies ($E = \frac{\hbar^2 k^2}{2m}$) associated with the corresponding values of k's. Such wave function describes a *wave packet* rather a plane wave. A wave packet is comprised of a group of waves of slightly varying wavelengths, whose amplitudes and phases are such that they interfere constructively over a small region of space, and outside of this region, they produce an amplitude that reduces to zero because of destructive interference.

Removing the time dependence from the previous equation, the following wavefunction is obtained:

$$\Psi(x, 0) = \frac{1}{\sqrt{2\pi}} \int_{-\infty}^{\infty} \emptyset(k) e^{i(kx)} dk \tag{4.12}$$

The Fourier transform of this equation is:

$$\emptyset(k) = \frac{1}{\sqrt{2\pi}} \int_{-\infty}^{\infty} \Psi(x, 0) e^{-i(kx)} dx \tag{4.13}$$

Thus, to derive the form of $\emptyset(k)$, it is necessary to know the initial wave function of the free quantum particle. The initial wave function $\Psi(x, 0)$ must be physically realizable to represent a real quantum particle. Thus, it is required that such a function be normalizable.

EXAMPLE 4.1 THE GAUSSIAN WAVE PACKET

Consider a free particle whose initial wave function is a Gaussian function and is given as:

$$\Psi(x, 0) = \alpha e^{-\gamma x^2} \tag{4.14}$$

α and γ are constants, and γ is positive and real. This initial wave function must be normalizable. The normalization of the previous function is:

$$1 = |\alpha|^2 \int_{-\infty}^{\infty} e^{-2\gamma x^2} dx \tag{4.15}$$

$$1 = |\alpha|^2 \sqrt{\frac{\pi}{2\gamma}} \rightarrow \alpha = \left(\frac{2\gamma}{\pi}\right)^{1/4} \tag{4.16}$$

Thus, the initial wave function is normalizable.

It is necessary to determine $\Psi(x,t)$ for the particle. Therefore, momentum-space wave function is determined by applying Equation (4.13):

$$\emptyset(k) = \frac{1}{\sqrt{2\pi}} \int_{-\infty}^{\infty} \alpha e^{-\gamma x^2} e^{-i(kx)} dx = \frac{\alpha}{\sqrt{2\pi}} \sqrt{\frac{\pi}{\gamma}} e^{-k^2/4\gamma} \tag{4.17}$$

Thus, a Gaussian function in x-space leads to a Gaussian function in k-space. A Gaussian function has a unique symmetry between x and k spaces. The wave function of the particle at any time "*t*" can now be written as:

$$\Psi(x, t) = \frac{1}{\sqrt{2\pi}}\left(\frac{2\gamma}{\pi}\right)^{1/4}\sqrt{\frac{\pi}{\gamma}}\int_{-\infty}^{\infty} e^{-k^2/4\gamma} e^{i(kx - \hbar\frac{k^2}{2m}t)}dk = \left(\frac{2\gamma}{\pi}\right)^{1/4}\frac{e^{-\gamma x^2/(1+i(\frac{2\hbar\gamma t}{m}))}}{\sqrt{1 + i\left(\frac{2\hbar\gamma t}{m}\right)}} \tag{4.18}$$

The probability density of the above wave function can therefore be simplified by describing as, $a = \frac{2\hbar\gamma t}{m}$, then

$$|\Psi(x, t)|^2 = \left(\frac{2\gamma}{\pi}\right)^{1/2}\frac{e^{-2\gamma x^2/(1+a^2)}}{\sqrt{1 + a^2}} \tag{4.19}$$

The graph of the above probability density function Equation (4.19) for any time $t > 0$ is a flattened Gaussian function. What does this mean? It means that as time passes, the probability density of the particle spreads out in space. This also implies that as time passes, the particle becomes delocalized and diffuses to several spatial points. Figure 4.1 illustrates such a change.

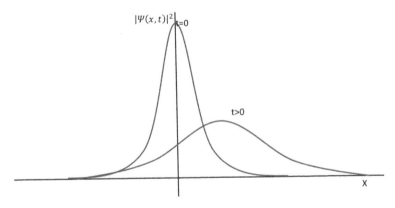

FIGURE 4.1 This figure illustrates the probability density spread at different times of the free particle whose wave function is a Gaussian function.

Conceptual Question 1: *The state vector of a spin-1/2 is described as a linear superposition of the discrete spin states. Here a free particle do not have discrete energy states. Does the principle of superposition still valid for the free particle? In which of the previous equations the principle of superposition is applied?*

A free classical particle moves with a constant velocity. What is the velocity of the quantum particle or wave packet?

To answer this question, it is important to understand what comprises a wave packet. It is comprised of individual waves that are superimposed and undergo constructive and destructive interference at different points in space. Therefore, this wave packet must have two velocities associated with it, the velocity of the individual waves and the velocity of the superimposed waves. The velocity of the individual waves is called the "phase velocity" and is described by the following relationship:

$$v_p = \frac{\omega}{k} \tag{4.20}$$

and

$$E = \hbar\omega = \frac{\hbar^2 k^2}{2m} \qquad (4.21)$$

Therefore,

$$v_p = \frac{\omega}{k} = \frac{E}{\hbar k} = \frac{\hbar k}{2m} = \frac{p}{2m} \qquad (4.22)$$

It seems that individual waves move with half of the classical velocity of the particle ($v_{classical} = \frac{P}{m}$).

The velocity of the superimposed waves or waves moving together as a group is called group velocity. It is given by the following relationship:

$$v_g = \frac{\partial\omega}{\partial k} \qquad (4.23)$$

Again since,

$$E = \hbar\omega \qquad (4.24)$$

and

$$\frac{\partial E}{\partial k} = \hbar\frac{\partial\omega}{\partial k} \qquad (4.25)$$

By substituting $E = \frac{\hbar^2 k^2}{2m}$, in the previous equation, the following group velocity is obtained:

$$v_g = \frac{\partial\omega}{\partial k} = \frac{\hbar k}{m} = \frac{p}{m} \qquad (4.26)$$

Thus,

$$v_g = \frac{p}{m} = v_{classical} = 2v_p \qquad (4.27)$$

TABLE 4.1
Summary of the Differences between Classical and Quantum Mechanical Particle

Free Classical Particle	Free Quantum Mechanical Particle
A point-like object that occupies a specific position in space. It moves in such a way that every portion moves in the same direction and at the same rate. It is simply represented as a dot in space.	A particle is an object that has plane waves associated with it. It is simply represented as a wave packet comprising of superimposed plane waves.
It moves in space as a localized object with a velocity that is given as $v_{classical} = \sqrt{\frac{2E}{m}}$	It has two velocities associated with it called the phase and group velocities. Phase velocity is half of the classical velocity of a particle, whereas group velocity is same as that of the classical velocity.
Free particles move continuously in space and time. Its state of motion remains constant if no forces are acting upon it.	The probability density of the particle changes with time even though no forces are acting on the particle. This is purely a quantum property.
It never spreads. The location of such a particle can be determined with certainty at any time.	It spreads with time. The uncertainty in position changes with time.
Its energy and momentum are constant in time.	Its energy and momentum are not constant in time, but its average energy and momentum remain constant with time.

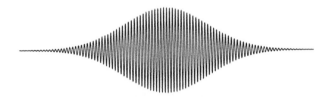

A whimsical depiction of a photon wave packet. It can be shown that for any shape of a normalized wave packet, the expectation value of the Hamiltonian, that is, the energy of the photon, is $\hbar\omega$, where $\hbar$ is the Planck constant divided by 2π and ω is the angular frequency associated with the wave vector of the photon.

FIGURE 4.2 An image of a photon wave packet from the following website https://www.nist.gov/image-21163, with permission from NIST. A photon is a quantum particle and can be represented by a wave packet. The ripples inside the wave packet travel with the phase velocity, and the wave packet (photon) moves with the group velocity.

The group of waves or simply a wave packet (quantum mechanical particle) moves with a velocity same as that of a classical free particle (Table 4.1 and Figure 4.2).

Conceptual Question 2: Is the state of a free particle considered as a stationary or nonstationary state?

PROBLEMS

4.1 A free particle has an initial wave function described by the following equation:

$$\Psi(x, 0) = \alpha e^{-\gamma \frac{|x|}{2}}$$

where γ and α are positive real constants. Normalize the wave function and find $\Psi(x, t)$ by using the above-mentioned method. Evaluate $\langle x \rangle$, $\langle p \rangle$, $\langle x^2 \rangle$, and $\langle p^2 \rangle$? Show that the Uncertainty principle is true at all times.

4.2 Evaluate the integral in Equation (4.18) and plot the probability densities at different times. Explain your observations from the probability density plot.

4.3 The solution of the Schrodinger equation for a free particle in one dimension consists of traveling waves moving to the right and to the left. Consider a free particle in three dimensions and write the Schrodinger equation for such a particle. Predict the solution of the Schrodinger.

4.4 What do you understand about the phrase "the spread of a wave packet with time." Explain. Draw diagrams to illustrate your explanation.

4.3 THE INFINITE SQUARE WELL

In the previous section, a quantum free particle is described as a wave packet traveling in space and time. Such a particle experiences no forces. In this section, a quantum particle under the influence of a force is discussed.

Consider the following potential energy function:

$$V(x) = \begin{cases} 0, & 0 \leq x \leq L \quad \text{(Region II)} \\ \infty, & \text{otherwise} \quad \text{(Region I and III)} \end{cases} \tag{4.28}$$

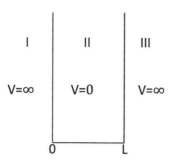

FIGURE 4.3   The infinite square well potential.

The infinite potential energy function in Region I and III confines the particle in Region II (well). The particle bounces back and forth between two ends ($x = 0$, $x = L$) of the well. There are no real potentials like this one that are known to exist. However, it may approximately represent some known potential energy functions. Why do we need to study such an ideal or unrealistic system? The mathematical simplicity of such an ideal system illustrates the concepts inherent in quantum behavior to allow the reader to better understand them in general. These concepts may be difficult to comprehend when considering a more realistic system.

In Section 4.2, it was discussed that a free particle has plane waves (also called as probability waves) associated with it. What will happen to the plane waves of the particle if you confine such a particle within a boundary? It will give rise to standing waves associated with a particle. It is informative to determine if the same conclusion can be derived by solving the Schrodinger equation for the above-mentioned system (Figure 4.3).

In Regions I and III, the potential is infinite, and thus, there exists no solutions to the Schrodinger equation. In Region II, the potential is zero. Applying the scheme of separation of variables discussed in Chapter 3, the time-independent Schrodinger equation for the previous boundary conditions can be solved. The time-independent Schrodinger equation for the particle in the infinite square well is defined by the following:

$$-\frac{\hbar^2}{2m}\frac{d^2\psi(x)}{dx^2} = E\psi$$

It can be written as:

$$\frac{d^2\psi(x)}{dx^2} = -k^2\psi(x), \ k^2 = \frac{2mE}{\hbar^2} \tag{4.29}$$

Since the potential inside the well is zero, the particle's energy is its kinetic energy, and $E \geq 0$. The general solution of the previous equation is:

$$\psi(x) = A \sin kx + B \cos kx \tag{4.30}$$

The potential function ($V(x) = 0$) is finite in region $0 \leq x \leq L$. However, for $x < 0$ and $x > L$, it is infinite. In such a case, the potential is said to have "hard walls" at $x = 0$ and $x = L$ (boundaries). Thus, the wave function must vanish for $x \geq L$, and $x \leq 0$. The derivative of the wave function (ψ') will be finite as $x \to 0$, $x \to L$. Thus, ψ' is discontinuous at the boundaries.

From these conditions, the following can be obtained:

$$\psi(x = 0) = \psi(x = L) = 0 \tag{4.31}$$

$$\psi(x = 0) = 0 = A \sin k(0) + B \cos k(0) \tag{4.32}$$

For Equation (4.32) to be true, it must be that $B = 0$. Therefore,

$$\psi(x) = A \sin kx \tag{4.33}$$

At the boundary $x = L$, $\psi(x = L) = 0 = A \sin kL$. For this to be satisfied, it is required that sin $kL = 0$, it follows therefore that:

$$kL = \pi, 2\pi, 3\pi, \ldots \tag{4.34}$$

Thus,

$$k_n = \frac{n\pi}{L}, \quad \text{where } n = 1, 2, 3, 4 \ldots \tag{4.35}$$

You may ask what about zero and negative values of number "n." All those features can be buried in the value of A. However, the value of the constant A still has not been determined. This constant can be determined by using the normalization condition for the wave function.

$$\int_0^L |\psi|^2 dx = \int_0^L |A|^2 \sin^2 kx \, dx = 1, \ |A|^2 = \frac{2}{L}.$$

Thus,

$$\psi_n(x) = \sqrt{\frac{2}{L}} \sin\left(\frac{n\pi}{L} x\right) \tag{4.36}$$

As is evident, "n" can have any positive value. Therefore, the solution of the Schrodinger equation for the previous system has an infinite number of solutions. These stationary states energy eigenfunctions are sinusoidal and similar to the standing waves depicted in Figure 4.4. Each solution corresponds to an energy eigenvalue. The particle has discrete energy values as shown:

$$k_n^2 = \frac{2mE_n}{\hbar^2} = \left(\frac{n\pi}{L}\right)^2$$

$$E_n = \frac{n^2\pi^2\hbar^2}{2\,mL^2} \tag{4.37}$$

The discretization of energy is the first quantum feature that emerges from this simple system. Can other quantum features be identified from this system? (Figure 4.4)

Simple Analogy: It is informative to think of an infinite square well as an infinitely tall, 2-dimensional building, whose width is "L" and has infinite numbers of floors. Each floor has associated with it a particular value of energy and corresponding energy eigenfunction. In this building, the lowest floor (ground floor) starts with $n = 1$ and not $n = 0$. At the lowest floor (E_1), the energy of the particle is at a minimum, and its energy eigenfunction is illustrated in the Figure 2, an "upside down hammock." In moving up the floors, the energy of the particle increases, and its energy eigenfunction is comprised of more than one "hammock," placed together in different configurations. However, the

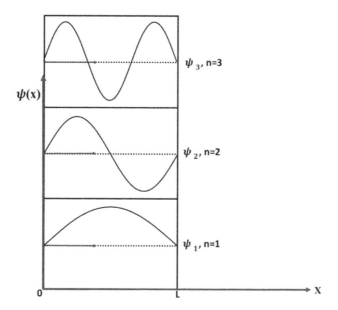

FIGURE 4.4 The wave functions of the first three stationary states are illustrated.

number of hammocks in each floor is equal to the floor number (n). So, at the ground floor, the energy eigenfunction is comprised of one hammock, at the second floor it will have two hammocks, one upside down, and so on. The particle is not resting at a particular location on a hammock but rather spreads out everywhere on these hammocks, except at the connecting points of the hammocks. This is because at these points $\psi(x) = 0$, and the particle's probability density is zero. These connecting points are called nodes. At the first floor, there are no nodes. At the second floor, there is one node point at $x = L/2$, and at the third floor, there are two node points, at $x = L/3$, and $x = 2L/3$. The number of node points increases with the floor number. The number of nodes for the nth level is $N = n - 1$.

All these eigenfunctions or stationary states must be mutually orthonormal and must satisfy completeness relationship.

   i. Orthonormality

$$\int \psi_m \psi_n^* \, dx = \delta_{mn} \tag{4.38}$$

  ii. Completeness

Any function $f(x)$ can be expressed as a linear combinations of the energy eigenfunctions,

$$f(x) = \sum\nolimits_{n=1}^{\infty} c_n \psi_n(x) \tag{4.39}$$

The coefficients c_n can be evaluated by evaluating the following integral. Multiply both sides of the previous equation by an energy eigenfunction as:

$$\int \psi_m(x) f(x) dx = \sum\nolimits_{n=1}^{\infty} c_n \int \psi_m(x) \psi_n(x) dx = \sum\nolimits_{n=1}^{\infty} c_n \delta_{nm} = c_m \tag{4.40}$$

$$\int \psi_m(x) f(x) dx = c_m \tag{4.41}$$

In the previous equation, generally $f(x)$ is the initial wave function $\psi(x, 0)$ of the system. The expansion coefficients c_n can be determined by using the initial wave function of a system. Without the knowledge of the initial wave function, it is not possible to determine these coefficients theoretically.

The two properties (orthonormality and completeness) of the stationary states are true for any quantum system and, in fact, form the basis for any acceptable solution of the Schrodinger equation.

The time-dependent part of the solution in the stationary state (as discussed in Chapter 3) will now be discussed. The time-dependent stationary state is:

$$\psi_n(x, t) = \sqrt{\frac{2}{L}}\sin\left(\frac{n\pi}{L}x\right)e^{-iE_nt/\hbar} = \sqrt{\frac{2}{L}}\sin\left(\frac{n\pi}{L}x\right)e^{-i\left(\frac{n^2\pi^2\hbar)t}{2mL^2}\right)} \tag{4.42}$$

The most general solution of the Schrodinger equation is a linear superposition of stationary states:

$$\Psi(x, t) = \sum_{n=1}^{\infty} c_n\sqrt{\frac{2}{L}}\sin\left(\frac{n\pi}{L}x\right)e^{-i\left(\frac{n^2\pi^2\hbar)t}{2mL^2}\right)} \tag{4.43}$$

EXAMPLE 4.2

The initial wave function ($\Psi(x, 0)$) of a particle in an infinite square well is in a linear superposition of two of the energy eigenfunctions or stationary states. What is the time-dependent wave function of the particle?

$$\Psi(x, 0) = c_1\psi_1(x) + c_2\psi_2(x) \tag{4.44}$$

Since c_1 and c_2 are complex numbers, the initial wave function can be written as:

$$\Psi(x, 0) = c(\psi_1(x) + e^{i\varphi}\psi_2(x)) \tag{4.45}$$

c is now a real number.

Normalizing the previous function to find c.

$$|\Psi|^2 = c^2(\psi_1(x)\,\psi_1(x)^* + \psi_1(x)\,\psi_2(x)^*e^{-i\varphi} + \psi_1(x)^*\psi_2(x)e^{i\varphi} + \psi_2(x)\psi_2(x)^*) \tag{4.46}$$

$$\int_0^L |\psi|^2 dx = 1 = c^2 \int_0^L (\psi_1(x)\,\psi_1(x)^* + \psi_1(x)\psi_2(x)^*e^{-i\varphi} + \psi_1(x)^*\psi_2(x)e^{i\varphi} + \psi_2(x)\psi_2(x)^*)dx$$

Since, the energy eigenfunctions are orthogonal, the middle terms will all vanish and

$$\int_0^L |\psi|^2 dx = 1 = c^2\left(\frac{2}{L}\right)\int_0^L\left[\left(\sin\left(\frac{\pi}{L}x\right)\right)^2 + \left(\sin\left(\frac{2\pi}{L}x\right)\right)^2\right]dx$$

$$1 = 2\,c^2$$

$$c = \frac{1}{\sqrt{2}}$$

Thus,

$$\Psi(x, 0) = \frac{1}{\sqrt{2}}(\psi_1(x) + e^{i\varphi}\psi_2(x)) \tag{4.47}$$

The time-dependent wave function of the particle is:

$$\Psi(x, t) = \frac{1}{\sqrt{2}}(\psi_1(x)e^{-iE_1 t} + e^{i\varphi}\psi_2(x)e^{-iE_2 t}) \qquad (4.48)$$

What is the expectation value of the particle's energy?

$$\langle \hat{H} \rangle = \langle \Psi(x, t)|\hat{H}|\Psi(x, t)\rangle = \frac{1}{2}\left[\left\langle \psi_1 \left| -\frac{\hbar^2}{2m}\frac{d^2}{dx^2} \right| \psi_1 \right\rangle + \left\langle \psi_2 \left| -\frac{\hbar^2}{2m}\frac{d^2}{dx^2} \right| \psi_2 \right\rangle\right] \qquad (4.49)$$

$$\langle \hat{H} \rangle = \frac{E_1 + E_2}{2} = \frac{5\pi^2 \hbar^2}{4\,mL^2} \qquad (4.50)$$

What is the expectation value of the particle's position? For simplicity let $\varphi = 0$
The particle's position expectation is:

$$\langle \hat{x} \rangle = \langle \Psi(x, t)|\hat{x}|\Psi(x, t)\rangle \qquad (4.51)$$

$$\Psi(x, t) = \frac{1}{\sqrt{2}}(\psi_1(x)e^{-iE_1 t} + \psi_2(x)e^{-iE_2 t})$$

And $E_n = n^2\,\omega$, $\quad \omega = \frac{\pi^2 \hbar^2}{2mL^2}$
Therefore:

$$\Psi(x, t) = \frac{1}{\sqrt{2}}(\psi_1(x)e^{-i\omega t} + \psi_2(x)e^{-i4\omega t}) \qquad (4.52)$$

$$\langle \hat{x} \rangle = \langle \Psi(x, t)|\hat{x}|\Psi(x, t)\rangle$$
$$= \frac{1}{2}\int_0^L \left[x\,\psi_1(x)\,\psi_1(x)^* + x\,\psi_2(x)\,\psi_2(x)^* + x(\psi_1(x)\psi_2(x)^* e^{i3\omega t} + \psi_1(x)^*\psi_2(x)e^{-i3\omega t})\right] dx \qquad (4.53)$$

$$\langle \hat{x} \rangle = \frac{1}{2}\int_0^L \left[x\sin^2\left(\frac{\pi x}{L}\right) + x\sin^2\left(\frac{2\pi x}{L}\right) + 2x\left(\sin\left(\frac{\pi x}{L}\right)\sin\left(\frac{2\pi x}{L}\right)\cos(3\omega t)\right)\right] dx \qquad (4.54)$$

$$\langle \hat{x} \rangle = \frac{L}{2}\left[1 - \frac{32}{9\pi^2}\cos(3\omega t)\right] \qquad (4.55)$$

By including the phase φ in the wave function, determine the difference in the expectation value of particle's position.

Conceptual Question 3: *What changes will you observe in the quantum properties of a free particle when it is confined in an infinite square well?*

PROBLEMS

4.5 What are the quantum properties of a particle in an infinite square well? To answer the question, consider a classical particle confined in an infinite square well. What will it do there? It will move back and forth within the boundaries, where its location can be determined exactly. Its total energy or simply kinetic energy can have only one fixed value. and it will follow Newton's laws of motion. Now try to answer this question based on the classical particle as a reference and identify the differences.

4.6 Consider a particle of mass "m" inside an infinite square well, having energy "E_n" as shown previously.

   We replace the particle inside the well by another particle of mass "$3m$." What would be the energy of the new particle? Now change the length of the well to $2L$. What would be the energy of the particle?

4.7 Find the solutions to the time-independent Schrodinger equation for the following delta function potential, $V(x) = \gamma \delta(x)$, where γ is a positive constant.

4.8 In Equation (4.54), the evaluation of the integrals is not shown. The first integral can be obtained in the following way:

$$\int_0^L x\sin^2\left(\frac{\pi x}{L}\right) dx = \int_0^L \frac{1}{2}\left[x - x\cos\left(\frac{2\pi x}{L}\right)\right] dx$$

$$\int_0^L \frac{1}{2}\left[x - x\cos\left(\frac{2\pi x}{L}\right)\right] dx = \left[\frac{x^2}{4} - \frac{x\sin\left(\frac{2\pi x}{L}\right)}{\frac{4\pi}{L}} - \frac{\cos\left(\frac{2\pi x}{L}\right)}{\frac{8\pi^2}{L^2}}\right]\Bigg|_0^L = \frac{L^2}{4}$$

Using the same approach obtain the second and third integrals.

4.9 Find the following expectation values, $\langle \hat{x} \rangle$, $\langle \hat{H} \rangle$, $\langle \hat{p} \rangle$, of a particle in an infinite square well where the particle's wave function is $\Psi(x, t) = \frac{1}{\sqrt{3}}\left(\psi_1(x)e^{-iE_1 t} + \psi_2(x)e^{-iE_2 t} + \psi_3(x)e^{-iE_3 t}\right)$.

4.10 Prove the orthonormality condition by evaluating the integral in Equation (4.44).

4.4 STEP POTENTIAL

In Section 4.2, a particle confined in an infinite square potential well was discussed. In this section, a free particle moving from left approaching a step potential as a barrier that has a constant value in a certain region is discussed. The form of function of the step potential is (Figure 4.5):

$$V(x) = \begin{cases} 0, & x < 0 \quad \text{(Region I)} \\ V_0 = \text{constant}, & x \geq 0 \quad \text{(Region II)} \end{cases}$$

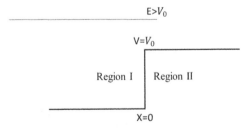

FIGURE 4.5   The step potential function. A particle approaches the step potential from the left. The energy of the particle is greater than the potential step.

Classically, if the energy "E" of the particle is greater than that of the potential barrier or potential step ($E > V_0$), the particle will penetrate the step potential region because it has enough energy to overcome the step potential barrier. It will slow down or speed up depending upon the potential barrier and keep moving to the right. Such a state of a particle is called a "scattering state."

A particle whose energy is less than the potential barrier ($E < V_0$) will turn around on reaching the barrier. The point at which it turns around is called as "turning point." It will never cross the barrier. Therefore, such a particle moving from the left will be reflected as it approaches the barrier.

A particle inside a barrier that has high potential walls such that the particle's energy (E) is much less than the height of wall of potential. Such a particle will be stuck between the walls and will oscillate back and forth between the walls (turning points). This state of the particle is called a "bound state."

Are there such states in quantum mechanical systems?

Yes, quantum mechanically such states do exist. For example, the state of a free particle that was studied in Section 4.1 is a scattering state. In such a case $V(x) = 0$ everywhere, there is nothing to bound such a particle, so it keeps moving to the right and goes back to infinity where it originated. If the particle is trapped inside the potential well, as in the case of infinite square well, the state of a particle is a bound state. In this case, the potential function is infinite, so the particle is trapped between the two walls of the barriers and can never escape.

Quantum mechanically, what can be predicted for a free particle moving from the left when it approaches a potential barrier? A free quantum particle is simply a wave packet; therefore, when it approaches the barrier from the left, a part of it will be reflected, and a part of it will be transmitted. However, we will not solve the Schrodinger equation for a wave packet approaching a potential barrier. Instead, we will adhere to treating the particle as a plane wave for mathematical simplicity. The same results (quantum properties) will be obtained. The Schrodinger equation for the particle in Region I and Region II can be solved, and it determines which quantum properties can be observed for the free quantum particle when it approaches the barrier.

Case I ($E > V_0$)

The Schrodinger equation for the particle moving from left in Region I (Figure 4.5) can be written as:

$$\frac{d^2 \psi_I(x)}{dx^2} = -k^2 \psi_I(x), \ k^2 = \frac{2mE}{\hbar^2} \tag{4.56}$$

and its solution is:

$$\psi_I(x) = A e^{ikx} + B e^{-ikx} \tag{4.57}$$

In the previous equation, the first term is an incident wave, and second term is a reflected wave. The probability density of the previous function remains constant with time, but it represents a particle moving to the right.

The Schrodinger equation for the particle in Region II is:

$$\frac{d^2 \psi_{II}(x)}{dx^2} = -p^2 \psi(x), \quad p^2 = \frac{2m(E - V_0)}{\hbar^2} \tag{4.58}$$

and its solution is:

$$\psi_{II}(x) = C e^{ipx} \tag{4.59}$$

Region I has incident and reflected waves, whereas Region II has only transmitted waves.

Applying condition of continuity of wave function and its derivative at the boundary ($x = 0$):

$$\psi_I(x = 0) = \psi_{II}(x = 0) \tag{4.60}$$

$$\frac{d\psi_I}{dx}\bigg|_{x=0} = \frac{d\psi_{II}}{dx}\bigg|_{x=0} \tag{4.61}$$

$$A + B = C \tag{4.62}$$

$$k(A - B) = pC \tag{4.63}$$

Solving the previous equations, the following amplitudes ratios are obtained:

$$\frac{C}{A} = \frac{2k}{k + p} \tag{4.64}$$

$$\frac{B}{A} = \frac{k - p}{k + p} \tag{4.65}$$

The previous equations are simply the amplitudes of reflected and transmitted waves, in terms of amplitude of the incident wave. The fraction of particles that are transmitted is equal to the ratio of transmitted current (here current means number of electrons or particle moving to the right) to incident current. The transmissivity is therefore:

$$T = \frac{|C|^2}{|A|^2}\frac{p}{k} = \frac{4pk}{(k + p)^2} \tag{4.66}$$

The reflectivity, R, is just the ratio of the intensities of the reflected and incident waves

$$R = \frac{|B|^2}{|A|^2} = \frac{(k - p)^2}{(k + p)^2} \tag{4.67}$$

The sum of reflectivity and transmissivity must be equal to 1. Thus,

$$T + R = 1 \tag{4.68}$$

From the previous equations, the reflectivity approaches zero, when p becomes equal to k. Therefore, $R = 0$, $E \sim E - V_0$, which means that when potential is negligibly small, reflection is almost zero. On the other hand, $R = 1$, when $p = 0$; that is, when E and V_0 are comparable in size, then all the particles are reflected. Except for these two conditions, the quantum particle will be reflected at the barrier no matter how much larger the energy of the particle is than the potential of the barrier.

Classically, as the particle approaches a barrier (potential), it slows down and crosses the barrier as long as its energy is greater than V_0 and will not be reflected. From the previous equations, quantum mechanically, the particle will be reflected no matter how much higher the energy of the particle is than the potential barrier. This reflection property of the particle when the energy of the particle is higher than the barrier is purely a quantum mechanical effect.

Case II ($E < V_0$)

In this case, a particle moving from the left encounters a barrier such that the energy of the particle is less than the potential barrier. In such a scenario, classically a particle will not be able to cross the

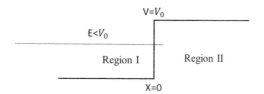

FIGURE 4.6   Step potential function (barrier) and a particle whose energy is $E < V_0$ approaches the barrier from the left.

barrier and will be completely reflected to the left. What is the quantum mechanical behavior of such a particle? (Figure 4.6)

$$V(x) = \begin{cases} 0, & x < 0 \quad \text{(Region I)} \\ V_0 = \text{constant}, & x \geq 0 \quad \text{(Region II)} \end{cases}$$

The Schrodinger equation remains the same, and only the value of "k" has changed. Thus, writing it again is a trivial step; rather, the solution of the Schrodinger equation is discussed.

In Region I, the solution of the Schrodinger equation gives the following wave function:

$$\psi(x) = Ae^{ikx} + Be^{-ikx} \tag{4.69}$$

$$k^2 = \frac{2mE}{\hbar^2}$$

In Region II, the solution of the Schrodinger equation yields the following wave function:

$$\psi(x) = Ce^{-lx} \tag{4.70}$$

where $l = \sqrt{\frac{2m(V_0 - E)}{\hbar^2}}$.

In this region, the wave function is exponentially decaying because $E < V_0$. Convince yourself of this by writing the Schrodinger equation.

Thus, there exists a solution of the Schrodinger equation in Region II, even when the energy of the particle is much less than the potential barrier. The wave function in Region II is an exponentially decaying function. Applying conditions of continuity (to the wave function and its derivative) at the boundary ($x = 0$) will yield the following amplitudes of the wave functions:

$$A = \frac{C}{2}\left(1 + i\sqrt{\frac{(V_0 - E)}{E}}\right) \tag{4.71}$$

$$B = \frac{C}{2}\left(1 - i\sqrt{\frac{(V_0 - E)}{E}}\right) \tag{4.72}$$

Therefore, the ratio of the reflected wave to that of the incident wave is

$$R = \frac{|B|^2}{|A|^2} = 1 \tag{4.73}$$

From the previous equation, it is clear that the quantum particle will be completely reflected. This result is in agreement with the classical prediction that we have already discussed (a particle cannot be transmitted if its energy is much less than the barrier).

Let

$$\sqrt{\frac{(V_0 - E)}{E}} = \tan(\varphi)$$

where φ = the phase of the wave function and substitute it in Equations (4.71) and (4.72). The following equation for the wave function in Region I is obtained:

$$\psi(x) = \frac{C}{2}\left(\frac{\cos\varphi + i\sin\varphi}{\cos\varphi}\right)e^{ikx} + \frac{C}{2}\left(\frac{\cos\varphi - i\sin\varphi}{\cos\varphi}\right)e^{-ikx} \tag{4.74}$$

Then, the wave function in Region I is

$$\psi(x) = L\cos(kx + \varphi) \tag{4.75}$$

where $L = \frac{C}{\cos(\varphi)}$ and is some new constant.

The wave function in Region I is oscillatory. However, the wave function in Region II is exponentially decaying. Also, the incident wave is completely reflected at the boundary of the barrier, as shown in Equation (4.73). What does this all mean?

First, the existence of exponentially decaying wave function of the particle in Region II, even though the energy of the particle is much less than the potential barrier, is purely a quantum mechanical effect. The implications of this is that although the incident particle is completely reflected at the boundary of the barrier, it will appear in Region II due to its wave-like properties. This implies that the particle can be found in Region II, where classically it is forbidden. Secondly, in Region II, the wave function of the particle is not oscillatory and is an exponentially decaying function. This implies that the likelihood of finding the particle is higher near the barrier than farther from it.

Figure 4.7 below describes the wave functions in Region I and II.

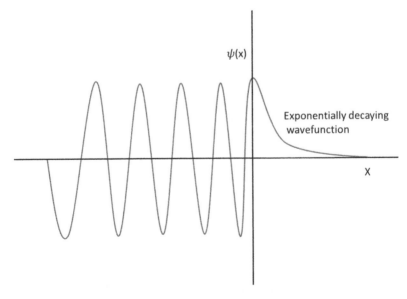

FIGURE 4.7 The visualization of the wave function, an oscillating wave function in Region I, and exponentially decaying function in Region II.

Conceptual Question 4: *Explain the differences between the quantum properties of Case I and Case II.*

Discussion: From the two previous cases, it is clear that quantum mechanical behavior of a particle defies all our classical intuitions of a particle or a wave that we developed over years. Why is the quantum particle completely reflected? Because it behaves like a classical particle. At the same time, why does it penetrate the classically forbidden region? The general notion is that a quantum particle has both particle and wave-like properties. It is due to wave-like properties that it penetrates Region II even when its energy is much less than the potential barrier. Why is the wave function not oscillatory in this region due to wave-like properties? This is because it behaves like a decaying wave near the boundary, rather like an oscillatory wave. Thus, the interaction of a particle with potential barrier (when the energy of the particle is less than the potential barrier) expresses the wave-like behavior of the particle. In simple words, we can say that the quantum particle behavior is complex. It does not fit into the classical criterion of a particle or a wave. Its behavior is dual (particle-wave), and some interactions bring out its wave-like properties, while some interactions bring out particle-like properties. This step potential is a good example of observing both particle-wave-like properties at the same time. Try explaining the previous two cases in your own words. Figure 4.8 illustrates the reflection and the penetration of a quantum particle (a wave packet) at a step potential.

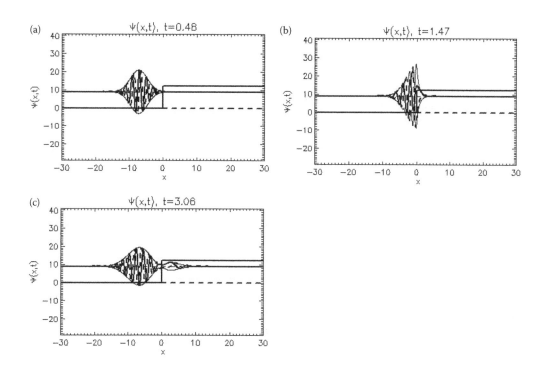

FIGURE 4.8   This figure illustrates the reflection of a wave packet from a step potential (a more realistic scenario). The image is taken from NIST website, https://www.ncnr.nist.gov/staff/dimeo/se_sim.html#Images with permission. It shows (a) an incoming wave packet approaching a step potential, (b) a wave packet being reflected and transmitted at the same time, and (c) a fully reflected and a partially transmitted wave packet.

PROBLEMS

4.11 By applying Equations (4.66) and (4.67), prove that $T + R = 1$.

4.12 For the Case II ($E < V_0$), write the Schrodinger equation for Region I and II, and determine the transmission coefficient.

4.13 Show that the solution of the Schrodinger equation in Region II gives an exponentially decaying wave function (Equation 4.70).

4.5 POTENTIAL BARRIER PENETRATION (TUNNELING)

In this section, a quantum particle moving from the left and approaching a square potential barrier is discussed, whose potential function is given as (Figure 4.9):

$$V(x) = \begin{cases} 0, |x < 0 & \text{(Region I)} \\ V_0, |0 \leq x \leq L & \text{(Region II)} \\ 0, x > L & \text{(Region III)} \end{cases}$$

Case I ($E < V_0$)

In Region I, the particle approaches the barrier from the left and is reflected at the boundary of the barrier. The solution of the Schrodinger equation in this region is a wave function that comprises both incident and reflected waves as shown in the following equation:

Region I

$$\psi_I(x) = Ae^{ik_1x} + Be^{-ik_1x} \tag{4.76}$$

where $k_1 = \sqrt{\frac{2mE}{\hbar^2}}$

In Region II, the particle penetrates the barrier. The solution of Schrodinger equation in this region gives the following wave function:

Region II

$$\psi_{II}(x) = Ce^{k_2x} + De^{-k_2x} \tag{4.77}$$

where $k_2 = \sqrt{\frac{2m(V_0-E)}{\hbar^2}}$

In Region III, the particle can escape from Region II and can be found in Region III. Thus, the solution of the Schrodinger equation in this region is the following wave function:

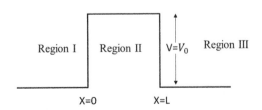

FIGURE 4.9 A schematic of the potential function showing Regions I, II and III.

Region III

$$\psi_{III}(x) = Fe^{ik_1 x} \tag{4.78}$$

By applying the conditions of continuity of the wave function and its derivative at the boundaries ($x = 0$, $x = L$), the following relationships between the coefficients can be obtained:

$$C = \frac{F}{2}\left(1 + i\frac{k_1}{k_2}\right)e^{(ik_1 - k_2)L} \tag{4.79}$$

$$D = \frac{F}{2}\left(1 - i\frac{k_1}{k_2}\right)e^{(ik_1 + k_2)L} \tag{4.80}$$

$$A = \frac{D}{2}\left(1 + i\frac{k_2}{k_1}\right) + \frac{C}{2}\left(1 - i\frac{k_2}{k_1}\right) \tag{4.81}$$

$$B = \frac{D}{2}\left(1 - i\frac{k_2}{k_1}\right) + \frac{C}{2}\left(1 + i\frac{k_2}{k_1}\right) \tag{4.82}$$

Therefore, the coefficient A in terms of coefficient F is:

$$A = \frac{F}{2}e^{ik_1 x}\left[\cos hk_2 L + i\frac{(k_2^2 - k_1^2)}{2k_2 k_1}\sin hk_2 L\right] \tag{4.83}$$

The transmission coefficient, which is the ratio of transmitted wave to that of incident wave is:

$$T = \frac{|F|^2}{|A|^2} = \frac{1}{\left[\cos h^2 k_2 L + \frac{(k_2^2 - k_1^2)^2}{(2k_2 k_1)^2}\sin h^2 k_2 L\right]} \tag{4.84}$$

The reflection coefficient is:

$$R = \frac{|B|^2}{|A|^2}$$

The previous equation for the transmission coefficient shows that there is a small probability that the particle is transmitted through the barrier or simply can penetrate the barrier. Classically, such a particle cannot penetrate the barrier. This property of the particle is purely a quantum mechanical effect. This barrier penetration is also called "tunneling." This phenomenon was first used to explain alpha particle decay of a nuclei. According to the theory of alpha decay, an alpha particle is retained inside the nucleus by the attractive nuclear forces. The energy of the particle is much less than the potential barrier. However, due to the wave nature of the particle, it can penetrate the barrier and can be found on the other side of the barrier. The probability that it can penetrate the barrier is given by Equation (4.84). In practical applications, the potential is not of rectangular form, as we have previously discussed. In Chapter 7, alpha decay will be discussed in more detail. An understanding of

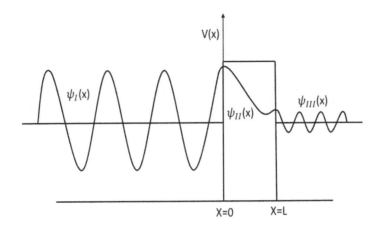

FIGURE 4.10 The wave function of the particle penetrating a square potential barrier. It is oscillatory function in Region I (comprised of incident and reflected wave), exponentially decaying function in Region II, and then oscillatory function again in Region III.

this phenomenon has been applied for building devices such as tunnel diode, the scanning tunneling microscope and tunnel junctions, which have become important components of the electronics industry (Figure 4.10).

PROBLEMS

4.14   Derive the expression for the reflection coefficient. Prove that $R + T = 1$.

4.15   Sketch the probability densities of the particle penetrating through the barrier in Region I, II and III.

4.16   Explain the difference between the barrier penetration of the particle in the case of step potential and the above potential.

4.17   Derive the expression for the transmission coefficient given in Equation (4.84) by using continuity conditions for the wave function and its derivative at the boundaries.

> **Conceptual Question 5:** *The penetration of a particle through the barrier is due to which quantum property?*

4.6 THE FINITE SQUARE WELL

In continuation of our study for the particle confined in a 1-dimensional space, this section begins with a problem of the particle under the influence of a finite 1-dimensional attractive potential. Consider that the space is divided into three regions, Region I, Region II and Region III. In each region, potentials are constant and have different values. The mathematical form of the potential function is described as (Figure 4.11):

$$V(x) = \begin{cases} 0, & x < -L \quad \text{(Region I)} \\ -V_0 = \text{constant}, & -L \leq x \leq L \quad \text{(Region II)} \\ 0 & x > L \quad \text{(Region III)} \end{cases}$$

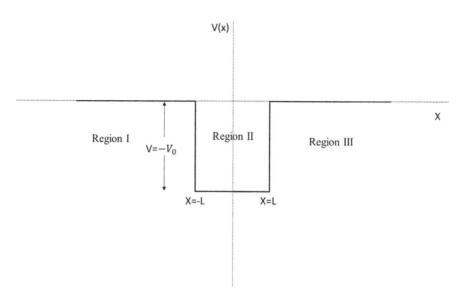

FIGURE 4.11   An attractive potential function.

The goal is to solve the Schrodinger equation in each region and find the corresponding wave function of the particle. This potential allows both scattering states ($E > 0$) and bound states ($E < 0$) as a solution. First, the bound states ($E < 0$) should be solved.

The Schrodinger equation in Region I is:

$$\frac{d^2\psi(x)}{dx^2} = -k^2\psi(x), \ k^2 = -\frac{2m|E|}{\hbar^2} \tag{4.85}$$

Since $E < 0$, the solution is an exponential function.
Its solution is:

$$\psi(x) = Ae^{kx} + Be^{-kx} \tag{4.86}$$

In this region, as $x \to -\infty$, the second term blows up. Therefore, the solution of the Schrodinger equation is:

$$\psi(x) = Ae^{kx} \tag{4.87}$$

Similarly, in Region III, the solution will remain same, but this time, the first term will blow up as $x \to \infty$; therefore, the solution is:

$$\psi(x) = Fe^{-kx} \tag{4.88}$$

In Region II, the Schrodinger equation is:

$$\frac{d^2\psi(x)}{dx^2} = -p^2\psi(x), \quad p^2 = \frac{2m(-|E| + V_0)}{\hbar^2} \tag{4.89}$$

and the solution is:

$$\psi(x) = Ce^{ipx} + De^{-ipx} \tag{4.90}$$

Solving for C and D, by applying continuity conditions at the boundary $(x = L)$ between Region II and Region III. The following coefficients are derived:

$$C = \frac{F}{2}\left(1 + i\frac{k}{p}\right)e^{-(k+ip)L} \tag{4.91}$$

$$D = \frac{F}{2}\left(1 - i\frac{k}{p}\right)e^{-(k-ip)L} \tag{4.92}$$

By applying the condition of continuity at the boundary $(x = -L)$ and solving for the coefficients, the following equations are derived:

$$-i\frac{k}{p} = \frac{Ce^{-ipL} - De^{ipL}}{Ce^{-ipL} + De^{ipL}} = \frac{\left(1 + i\frac{k}{p}\right)e^{-(2ip)L} - \left(1 - i\frac{k}{p}\right)e^{(2ip)L}}{\left(1 + i\frac{k}{p}\right)e^{-(2ip)L} + \left(1 - i\frac{k}{p}\right)e^{(2ip)L}} \tag{4.93}$$

To simplify the above expression, let:

$$\frac{k}{p} = \tan(\varphi), \quad 1 + i\frac{k}{p} = \frac{e^{i\varphi}}{\cos(\varphi)}$$

Now, plug in the simplification terms into Equation (4.93). The following is obtained:

$$\tan(\varphi) = \tan(2pL - \varphi) \tag{4.94}$$

The previous equation implies that $\varphi = 2pL - \varphi + N\pi$, where N is any integer. Thus

$$\varphi = pL + \frac{N\pi}{2} \tag{4.95}$$

$$\tan(\varphi) = \begin{cases} \tan(pL), & |N \text{ even} \\ -\cot(pL), & |N \text{ odd} \end{cases} \tag{4.96}$$

Expressing the previous equation in terms of k and p yields the following:

$$\frac{\sqrt{|E|}}{\sqrt{V_0 - |E|}} = \begin{cases} \tan\left(\dfrac{L\sqrt{2m(V_0 - |E|)}}{\hbar}\right), & |N \text{ even} \\[3mm] -\cot\left(\dfrac{L\sqrt{2m(V_0 - |E|)}}{\hbar}\right), & |N \text{ odd} \end{cases} \tag{4.97}$$

The previous equation is a transcendental equation for the energy. The solution of the equation will yield the possible energy levels. The equation can only be solved numerically or graphically. To do this, the equations must be rewritten with the following substitutions.

Let

$$z = \frac{L\sqrt{2m(V_0 - |E|)}}{\hbar} \tag{4.98}$$

$$z_0 = \frac{L\sqrt{2mV_0}}{\hbar} \tag{4.99}$$

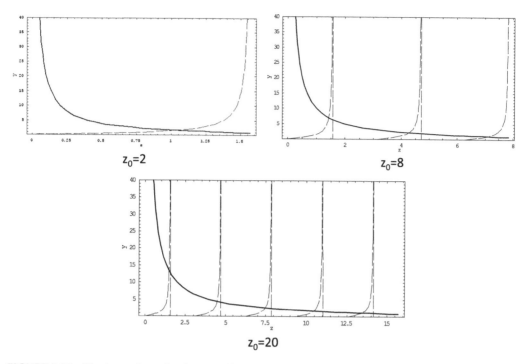

$z_0=2$

$z_0=8$

$z_0=20$

FIGURE 4.12   The three schematics show graphical solution of Equation (4.101) for different value of z_0 ($z_0 = 2$, $z_0 = 8$, and $z_0 = 20$) with bigger value of z_0 corresponding to a deeper well.

These yields:

$$\frac{\sqrt{|E|}}{\sqrt{V_0 - |E|}} = \frac{\sqrt{z_0^2 - z^2}}{z} = \sqrt{\frac{z_0^2}{z^2} - 1} \qquad (4.100)$$

For N even, the following relationship is obtained:

$$\tan(z) = \sqrt{\frac{z_0^2}{z^2} - 1} \qquad (4.101)$$

Graphically plotting the functions $y_1 = \tan(z)$, and $y_2 = \sqrt{\frac{z_0^2}{z^2} - 1}$ on the same grid will provide the bound state for the potential function. The intersection points of these functions will give the bound state energy levels. The Figure 4.12 shows the plots of these functions on the same grid for $z_0 = 2$ and $z_0 = 8$ and $z_0 = 20$. The plots show that for a shallow well ($z_0 = 2$) the number of bound states is one only, and as the potential well is made deeper by increasing $z_0 = 8$, and $z_0 = 20$, the number of bound states increases. Thus, for any finite value of V_0, there is only a finite number of bound states. For a very shallow well, no matter how shallow it is, there is always one bound state (even states).

Conceptual Question 6: Explain the differences between the quantum properties of infinite square well and finite square well.

PROBLEMS

4.18 For the previous potential well, derive the expressions for the transmission and reflection coefficients for scattering states ($E > 0$).

4.19 Derive the transcendental equation when N is odd and solve graphically. What is the difference between the energy values for N odd and N even? Do the same analysis (for a shallow well corresponding to even states has always at least one bound state, does the same apply for shallow well for odd states) as above is true for very deep and shallow well for odd values of N.

4.20 All attractive potentials in 1 dimension have at least one bound state. Prove this theorem.

4.21 A particle of mass "m" is approaching a potential well, as shown in Figures 4.13(a) and (b): ("E" is total energy of the particle, V_0 is the potential energy).

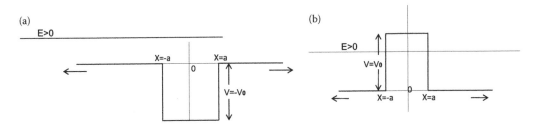

FIGURE 4.13

Write the Schrodinger equation for both potential barriers. Predict the quantum behavior of the particle that will be observed without solving the Schrodinger equation.

4.22 Explain what is quantum mechanical tunneling? Give a physical example. Also, explain why tunneling is not allowed in classical physics?

4.23 Consider the following potential wells: Figure 4.14 (V_0 is the potential energy)

Now, explain in which of the these two scenarios a particle of mass "m" and total energy "E" will be in the bound state? Write the wave function of the particle outside and inside the well for both cases.

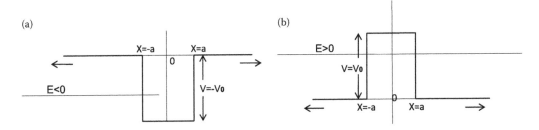

FIGURE 4.14

4.24 Consider that a particle is tunneling through a potential barrier, as shown in Figure 4.9. Does the total energy of the particle change as it passes from Region I to II and then to III? What is the total energy of the particle in Region III? Assume that initially the total energy of the particle in Region I was "E."

4.7 TUTORIALS

4.7.1 INFINITE SQUARE WELL

Purpose

To understand the behavior of a quantum particle under the influence of infinite square-well potential.

Concepts

- An infinite square well represents a system where the potential is infinite everywhere except for a small region (where it is zero).
- A particle in such a system is confined to move freely only in a region where the potential is zero. The force experienced by the particle in this region is also zero, and everywhere outside the region of confinement is infinite. This scenario is also called as "particle in a box."
- The total energy of the particle is only kinetic energy and is positive.
- The solution of the Schrodinger equation of the particle gives infinite discrete energies.

Consider Figure 4.15 given below, a particle confined in a region of length "a," called region II.

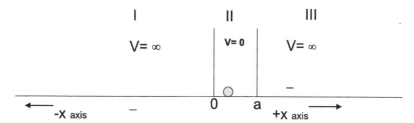

FIGURE 4.15   The figure is an illustration of infinite square well.

Question 1   In which region is the particle's wave function non-zero? Is the wave function finite/ infinite or zero in regions I and III? Explain. What should be the wave function at the boundaries ($x = 0$, and $x = a$)? Explain the meaning of "hard walls."

_____

_____

_____

_____

_____

Question 2   Sketch any mathematical function that can satisfy the previous boundary conditions for the particle in an infinite square well.

_____

_____

_____

_____

Question 3   Assume that the potential function in the previous figure is finite. Write the appropriate equations for continuity of the wave function and its derivative at the boundaries for a quantum particle.

_____

_____

_____

_____

_____

The Schrodinger Equation

The time-independent Schrodinger equation of a particle in a 1-dimensional system is the following:

$$\frac{-\hbar^2}{2m}\frac{d^2\psi}{dx^2} + V(x) = E\psi$$

Question 4   Write the Schrodinger equation for the particle in region II.

_____

_____

_____

_____

_____

Wave function and energy states

The solution of the Schrödinger equation for the previous system gives an infinite set of orthogonal energy eigenfunctions, and these functions are:

$$\psi_n = \sqrt{\frac{2}{a}}\sin\left(\frac{n\pi}{a}x\right)$$

where $n = 1, 2, 3\ldots$

The energy eigenvalue associated with each energy eigen function (stationary state) is:

$$E_n = \frac{n^2\pi^2\hbar^2}{2ma^2}$$

The most general solution of the Schrodinger equation is:

$$\psi(x) = \sqrt{\frac{2}{a}}\sum_{n=1}^{\infty} c_n \sin\left(\frac{n\pi}{a}x\right)$$

where c_n is the expansion coefficient.

Question 5   Explain the meaning of the term, $|c_n|^2$.

_____

_____

_____

_____

_____

Question 6   Explain the difference in the meanings of the terms, $|\psi(x=a)|^2$, and $|\psi(x)|^2dx$.

_____

_____

_____

_____

_____

It is crucial to learn how to calculate expansion coefficients.

To calculate these coefficients, one should know the wave function at $t = 0$. The expansion coefficient can be calculated as:

$$c_n = \sqrt{\frac{a}{2}} \int_0^a \sin\left(\frac{n\pi}{a}x\right)\psi(x)dx$$

Question 7   If the initial wave function of the particle is $\psi(x, 0) = A$, where "A" is arbitrary constant, determine the expansion coefficients.

_____

_____

_____

_____

_____

Consider that the particle initially was in an energy eigenstate (E_n) given as follows and its wave function evolves with time.

$$\psi_n = \sqrt{\frac{2}{a}}\sin\left(\frac{n\pi}{a}x\right)$$

Question 8   What is the energy of the particle at a later time "t"? Is there any difference between the energy of the particle at $t = 0$ and at $t = t$? Also, determine the wave function of the particle at later time "t."

_____

_____

_____

_____

_____

Consider that the particle starts initially $(t = 0)$ in a linear superposition state as follows:

$$\psi(x) = \frac{1}{\sqrt{2}}\psi_1(x) + \frac{1}{\sqrt{2}}\psi_2(x)$$

Question 9   Determine the wave function of the particle at later time "t." What will be the energy of the particle at a later time?

_____

_____

_____

_____

_____

Question 10   If the particle is in an energy eigenstate $n = 1$, find the following expectation values:

$$\langle x \rangle, \langle p \rangle, \langle p^2 \rangle, \langle H \rangle.$$

_____

_____

_____

_____

_____

Also, for the linear superposition state as shown previously, calculate the same expectation values. What difference do you find in the values for both cases?

Question 11 Plot the eigenfunctions for $n = 1, n = 2, n = 3, n = 10, n = 100$ and interpret their differences. Also, plot the probability density of the particle for the eigenstates $n = 1, n = 10$, and $n = 100$.

Question 12 Find the probability density of the particle at location $a/2$, assuming that the particle's energy is E_1. Also, do the same when the particle is in the previous linear superposition state.

Question 13 What happens to the wave function of the particle after a measurement of any quantity such as energy, position or momentum is performed?

Question 14 Find the energy eigenfunctions of the particle by changing the region of confinement from "0 to $-a$," "$-a$ to a" and "0 to $2a$" without solving the Schrodinger equation. Is there any change in the characteristic behavior of the energy eigenfunctions due to change in the region of confinement?

Question 15   List the physical quantities that can be calculated using the wave function and interpret its usefulness.

_____

_____

_____

_____

_____

4.7.2   Tunneling

Purpose

To help students understand the classical rules for the penetration of a particle and a wave through the barrier.

   Qualitative introduction to quantum tunneling for lower-level and upper-level undergraduate classes.

Concepts

- Classically, a particle can penetrate through the barrier only when the energy of the particle is greater than the potential energy of the barrier.
- The energy of the particle as it penetrates from one region to another does not change.
- Classically, kinetic energy of the particle can never become negative.
- Waves are spread out in space (delocalized) and, therefore, can penetrate through the barrier even when the energy of the wave is less than the potential energy of the barrier.
- Quantum mechanically, a particle has both wave and particle-like properties. Therefore, it can penetrate through the barrier even when its energy is less than the energy of the barrier because of its wave-like behavior.

Let us consider a particle of mass "m" and total energy "E" moving from left as shown in the Figures 4.16a and 4.16b that follow.

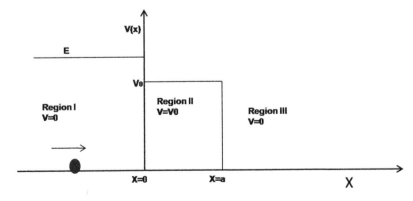

FIGURE 4.16   (a) $E > V_0$, the rectangle shown is called the "rectangular potential barrier."

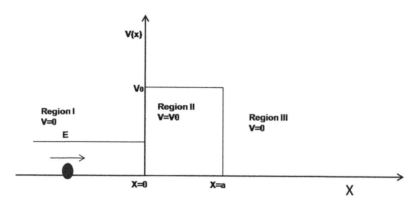

FIGURE 4.16 (b) $E < V_0$.

Question 1 In which of the two scenarios shown previously is the particle classically allowed to penetrate through the barrier?

_____

_____

_____

_____

_____

Question 2 On the basis of your previous answer, explain why the particle is not classically allowed to penetrate into region II in one of the scenarios.

_____

_____

_____

_____

Question 3 What is the region where a particle can never be found classically?

_____

_____

_____

_____

_____

Question 4 In which of the two scenarios shown previously does the particle slow down in region II?

Prove it. Hint: $E = P.E + K.E$

_____

_____

_____

_____

_____

Let us now replace the particle with a traveling wave (light wave or microwave) as shown in the Figures 4.16c and 4.16d that follows.

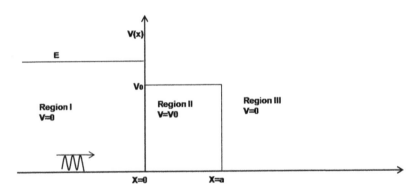

FIGURE 4.16   (c)

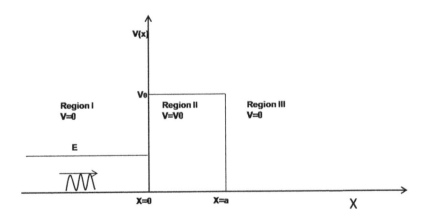

FIGURE 4.16   (d)

Question 5 In *both* scenarios shown previously, the traveling wave is classically allowed to penetrate through the barrier. Explain why.

_____

_____

_____

_____

_____

Question 6 What do you think will happen to the wave (whether it gets transmitted/reflected) in Figures 4.16c and 4.16d, when it encounters the boundary of region II?

_____

_____

_____

_____

_____

Question 7 Consider Figure 4.16d and answer the following question.
"The kinetic energy of the wave can never become negative in region II as long as its total energy is positive." Do you agree/disagree with this statement? Explain why you agree or disagree with the statement.

_____

_____

_____

_____

_____

Quantum Tunneling

Since you understand the classical rules for penetration of a particle and a wave through the barrier, now it will be easy to understand quantum mechanical tunneling.

Let us now consider a particle of mass "m," and energy "E" is entering from left (region I) as shown in the Figure 4.16e that follows. Such a particle has both wave and particle-like properties.

When such a particle that has both wave and particle-like properties encounters the boundary of region II, then just like a wave, the particle has a finite probability of being reflected and transmitted across the boundary.

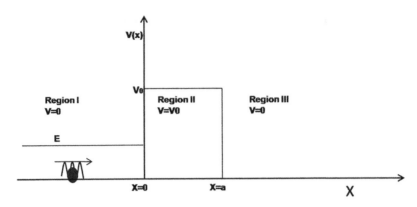

FIGURE 4.16   (e)

Question 8   Based on the concepts that you learned in the previous part of the tutorial, now make a guess whether a quantum particle can penetrate through the barrier as shown in Figure 4.16e. If you think a quantum particle can penetrate through the barrier, explain which characteristic property of the particle is responsible for the penetration of the particle through the barrier.

_____

_____

_____

_____

_____

The potential energy and wave functions of the particle are shown in the table that follows:

	Region I ($x < 0$)	Region II ($0 \leq x \leq a$)	Region II ($x > a$)
$V_0(x)$, Potential Energy	0	V_0	0
Time-independent Schrodinger equation		$\frac{d^2\Psi_2}{dx^2} - \frac{2m}{\hbar^2}(V_0 - E)\Psi_2 = 0$	
Wave number	$k_1 =$	$k_2 = \sqrt{2m(V_0 - E)/\hbar^2}$	$k_3 =$
$\Psi(x)$, Wave function	$\Psi_1 = Ae^{ik_1 x} + Be^{-ik_1 x}$ $\quad\downarrow\qquad\downarrow$ $\quad I\qquad R$	$\Psi_2 = ?$	$\Psi_3 =$ $\downarrow$ T

I: Incident wave
T: Transmitted wave
R: Reflected wave

Fill out the table above and write the Schrodinger equation in regions I and III and the form of the wave function and wave numbers.

Question 9 Let us assume that there is no reflection in region II. Write the wave function. Is this wave function exponentially decaying? Show it.

_____

_____

_____

_____

Question 10 Define transmission probability. Write the expression for transmission probability in terms of incident and transmitted wave.

_____

_____

_____

_____

Question 11 "A traveling wave is continuous between the boundary of two regions; therefore, its wave function and the derivative of the wave function should also be continuous."
Do you agree/disagree with the previous statement? If you agree, please write this statement mathematically at the boundary of regions I and II and regions II and III as shown in the previous figure.

_____

_____

_____

_____

Now listen to the following conversation between two students:

Student A: The quantum particle dissipates energy as it penetrates through the barrier, and, therefore, the wave function decays exponentially inside the barrier. This is similar to what we have learned in electromagnetism class that when an electromagnetic wave passes through a conductor it dissipates energy and dies out after traveling a certain length.

Student B: I do not think so because there is no term in the time-independent Schrodinger equation that is responsible for the dissipation of energy. I think the total energy of the particle is conserved and remains the same in all the regions. For the particle to dissipate energy, it should interact with something.

Question 12   Do you agree with Student A or Student B? Explain why you agree.

_____

_____

_____

_____

_____

Question 13   Let us consider that the quantum particle is traveling from the left as shown in Figure 4.16e. Under the condition that $K\,a >>> 1$, where $K = \sqrt{2\,m\,(V_0 - E)/\hbar^2}$ and a is the width of the barrier, then the transmission probability becomes proportional to the following function and is given as:

$$T\;\alpha\;e^{-2\,K\,a} \qquad\qquad\qquad (4.102)$$

Now, if the height and the width of the potential barrier is made twice, does the transmission probability shown increase or decrease? Plot the transmission function probability between $x = 0$ and $x = a$.

_____

_____

_____

_____

_____

Question 14   Transmission probability depends on which parameters? List them. (Use the previous relationship to answer your question.)

_____

_____

_____

_____

_____

Simulation

Use the following simulation for enhancing your understanding of quantum mechanical tunneling.

http://phet.colorado.edu/simulations/sims.php?sim=Quantum_Tunneling_and_Wave_Packets
Click on Run.

On the right side of the figure, select *barrier/well* from the top, and then select for *show energy values, show reflection and transmission probabilities, plane waves* and for "real part."

Now, on the top of the screen there is an option for "configure energy." Click it. A window will appear. Now, vary "average total energy," potential energy and the width of the barrier with the values of energies shown in the table that follows and run the simulation. Always keep the barrier position, $B1 = 0$. The average total energy is the energy of the plane wave, and the potential energy is the height of the barrier. On the screen, you will see display of T (transmission probability) and R (reflection probability).

Fill out the following table and record your observations:

$V_1 = 0$ $V_2 = 0.40$ eV $V_3 = 0$ Barrier Width $B1 = 1$ nm	Average Total Energy (E) (eV)	Transmission Probability (T)
	0.05	
	0.15	
	0.20	
	0.25	
	0.35	
	0.40	
	0.45	
	0.55	
	0.65	

$V_1 = 0$ $V_2 = 0.80$ eV $V_3 = 0$ Barrier Width $B1 = 1$ nm	Average Total Energy (E) (eV)	Transmission Probability (T)
	0.15	
	0.25	
	0.45	
	0.55	
	0.75	
	0.80	
	0.85	
	0.95	

$V_1 = 0$ $V_2 = 0.80$ eV $V_3 = 0$ Barrier Width $B1 = 2.0$ nm	Average Total Energy (E) (eV)	Transmission Probability (T)
	0.15	
	0.25	
	0.45	
	0.55	
	0.75	
	0.80	
	0.85	
	0.95	

Question 15 Describe the trends that you have observed for average total energy versus transmission probability in Tables I and II, such as when $E \ll V_2$, $E \gg V_2$ and $E \approx V_2$. Sketch a graph between transmission probability (T) versus energy (E) for these tables. Is there any similarity between these two graphs? Do these graphs validate the relationship between transmission probability and energy as shown above in Equation (4.102)?

_____

_____

_____

_____

_____

Describe your observations for Tables II and III when you increased the width of the barrier by a factor of two, keeping the height of the barrier same.

_____

_____

_____

_____

_____

Do the same for the wave packet.

_____

_____

_____

_____

_____

Question 16 Explain in your own words: What is tunneling? Also explain: Why, quantum mechanically, is a particle allowed to penetrate a classically forbidden region?

_____

_____

_____

_____

_____

$V_1 = 0$ $V_2 = 0.40$ eV $V_3 = 0$ Barrier Width $B1 = 1$ nm	Average Total Energy (E) (eV)	Transmission Probability (T)
	0.05	0.0
	0.15	0.02
	0.20	0.04
	0.25	0.07
	0.35	0.18
	0.40	0.28
	0.45	0.40
	0.55	0.71
	0.65	0.93

$V_1 = 0$ $V_2 = 0.80$ eV $V_3 = 0$ Barrier Width $B1 = 1$ nm	Average Total Energy (E) (eV)	Transmission Probability (T)
	0.15	0.0
	0.25	0.0
	0.45	0.01
	0.55	0.02
	0.75	0.11
	0.80	0.16
	0.85	0.24
	0.95	0.52

$V_1 = 0$ $V_2 = 0.80$ eV $V_3 = 0$ Barrier Width $B1 = 2.0$ nm	Average Total Energy (E) (eV)	Transmission Probability (T)
	0.15	0.0
	0.25	0.0
	0.45	0.0
	0.55	0.0
	0.75	0.01
	0.80	0.05
	0.85	0.32
	0.95	0.62

4.7.3 Quantum Wave Packet

Purpose

To help students understand the concept of wave packet of a quantum particle.

Concepts

- A wave packet of a quantum particle gives the information where the particle is most likely or unlikely to be found in space at any instant of time.
- Classically, group velocity of a wave packet may be greater or less than the phase velocity. When the phase velocity of all the waves is same as the group velocity, then the wave packet

does not experience any dispersion, and the medium through which it travels is called as nondispersive medium. A vacuum is a nondispersive medium for electromagnetic waves, whereas water acts as a dispersive medium for water waves.

- A wave packet of a free quantum particle experiences dispersion irrespective of the medium.
- The group velocity of a wave packet of a free quantum particle is same as the velocity of a free classical particle.

Question 1   Which of the following statements are true?

    I. An electron that has both wave and particle-like behavior and is completely delocalized in space can be represented by a sinusoidal wave.

    II. An electron that has both wave and particle-like behavior and is completely delocalized in space cannot be represented by a sinusoidal wave.

    III. An electron that has both wave and particle-like behavior and is completely delocalized in space can be represented by a wave packet.

    IV. An electron that has both wave and particle-like behavior and is completely delocalized in space cannot be represented by a wave packet.

      a. I and III
      b. I and IV
      c. I only
      d. II only
      e. None of the above

Question 2   Which of the following statements are true?

    I. An electron that has both wave and particle-like behavior and is localized in some region of space can be represented by a sinusoidal wave.

    II. An electron that has both wave and particle-like behavior and is localized in some region of space cannot be represented by a sinusoidal wave.

    III. An electron that has both wave and particle-like behavior and is localized in some region of space can be represented by a wave packet.

    IV. An electron that has both wave and particle-like behavior and is localized in some region of space cannot be represented by a wave packet.

      a. I and III
      b. I and IV
      c. III only
      d. II only
      e. None of the above

Question 3

Now, listen to the following conservation between two students who are discussing why an electron can be represented as a wave packet and answer the question that follows:

Classico: An electron has wave-like behavior, and, therefore, when an electron behaves like a wave, it can be represented as a wave packet.

Quantix: An electron can be represented as a wave packet because of the superposition principle, as quantum mechanically, an electron is represented as a linear superposition of sinusoidal waves of different frequencies, and this linear superposition of waves forms a wave packet. Moreover, physically, an electron is localized in space, which means that the probability of finding the electron in some region of space is non-zero, whereas in other regions, this probability is zero. Since a wave packet is like an envelope, and if we assume that the electron is inside this envelope and moves in space along with this envelope, then it is almost the same as saying that the electron is localized in a region of space (envelope) where the probability of finding the electron in that region (envelope) is non-zero, whereas outside of that region electron can never be found (probability of finding the electron is zero). Therefore, an electron can be represented as a wave packet.

Classico: If the electron is localized in space, then we always know where the electron is at any instant of time. Then why do we need to represent it as a wave packet and not simply as a particle?

Quantix: Because the electron has wave-like behavior and is spread out everywhere in the region of space where it is localized. Even if we know where it is located at any instant of time, we do not know what the momentum of the electron at that location is. Therefore, if we don't know its momentum at that location, then we cannot represent it as a particle.

Explain why you agree or disagree with each student.

_____

_____

_____

_____

_____

Question 4   Explain, in your own words, why a quantum particle can be represented by a wave packet.

_____

_____

_____

_____

_____

Question 5   The velocity with which the wave packet of a free quantum particle travels is

_____

_____

_____

_____

_____

Question 6   The group and phase velocity of a free quantum particle in terms of its wavelength and wave number is

_____

_____

_____

_____

_____

Question 7   Suppose we plot the square of the function of wave packet in position space. The area under this function is equivalent

_____

_____

_____

_____

_____

Question 8   Consider the following dispersion relationship of a free quantum particle:

$$\omega = \frac{hk^2}{2\pi m} \tag{4.103}$$

Now, use Equation (4.104) that follows to show that the wave packet of quantum particle experiences dispersion.

$$\omega(k - k_0) = \omega(k_0) + (k - k_0)\frac{d\omega}{dk}\bigg|_{k=k_0} + \frac{(k - k_0)^2}{2}\frac{d^2\omega}{dk^2}\bigg|_{k=k_0} + \cdots \tag{4.104}$$

_____

_____

_____

_____

_____

Question 9   Describe the difficulties and contradictions that arise when a free quantum particle is represented as a sinusoidal wave (or a plane wave). Also, explain how these contradictions are resolved when a quantum particle is instead represented as a wave packet.

_____

_____

_____

_____

_____

Question 10   Describe what happens to the wave packet when a measurement is performed to determine its position.

_____

_____

_____

_____

_____

Question 11   Consider the following expression for the quantum wave packet:

$$f(x, t) = \int dk\, g(k)e^{i(kx-\omega t)}$$

where $g(k) = e^{-a(k-k_0)^2}$

Now, derive the expression for the absolute square of this function. What is the physical meaning of the absolute square of function? Also, show that the width of the wave packet varies with time.

_____

_____

_____

_____

_____

Question 12 Write the fundamental differences between classical and quantum wave packet.

_____

_____

_____

_____

_____

Simulation

Use the following simulation to study the dispersion of a wave packet:

http://www.falstad.com/dispersion/

In this simulation, two sinusoidal waves of different frequencies are superimposed to produce a wave packet at the bottom. Follow the instructions and the answer the questions.

1. Make Speed 1 = Speed 2, and slowly slide across the bar of frequency 1 while maintaining frequency 2 as constant.

   Question 13 What happens to the width of the wave packet when the frequency difference between wave 1 and 2 is large? Draw the pattern that you observe. Should the difference between the frequencies of wave 1 and 2 be very small (narrow frequency range) to observe a wave packet of large width? Draw the pattern of the wave packet when the difference between the frequencies of wave 1 and 2 is very small.

   _____

   _____

   _____

   _____

   _____

Question 14   Should the width of the wave packet be small or large for the quantum particle to be well localized?

_____

_____

_____

_____

_____

Question 15   What can you tell about the momentum of the particle whose wave packet width is very large? You can try making the width of the wave packet very large.

_____

_____

_____

_____

_____

Question 16   Under above conditions (Speed1 = Speed2), do you observe dispersion of the wave packet?

_____

_____

_____

_____

_____

Question 17   Find the conditions (speed, frequency) under which the shape of the wave packet changes maximally with time.

2. Listen to the following conversation between three students discussing the dispersion of a free quantum particle.

> _Student A_: The wave packet of a free quantum particle disperses because its shape keeps changing with time.
> _Student B_: Um…, Wave packets do not disperse as the total probability density of finding the particle is constant.
> _Student C_: The wave packet disperses because the second derivative in the dispersion relation is not zero, and phase and group velocities are not same.

With which student(s), if any do you agree? Explain your reasoning.

_____

_____

_____

_____

_____

5 Applications of the Formalism-II

5.1 THE HARMONIC OSCILLATOR

A harmonic oscillator is a system that exhibits simple harmonic motion or periodic motion, a motion that repeats itself after equal intervals of time. Such a motion is caused by a restoring force, which is proportional to displacement of the oscillator and acts in a direction opposite to the displacement. Many systems such as spring, simple pendulum, vibrating string and molecular vibration can be approximated as simple harmonic oscillators. In classical physics, such systems are well understood. Therefore, the most natural question to ask is, what is a quantum mechanical simple harmonic motion? What are its properties? Its study is key to understanding the vibration of individual atoms in molecules and crystals. It is also very important for understanding particle properties of an electromagnetic wave. In this chapter, first we solve the Schrodinger equation for a harmonic oscillator using two methods, analytical method and algebraic method. Later, we discuss the quantum properties of the harmonic oscillator.

5.1.1 ANALYTICAL METHOD

We begin this chapter by considering a simple harmonic oscillator, a spring whose spring constant is "k" attached to a mass "m" as a simple harmonic oscillator. The mathematical and graphical form (Figure 5.1)

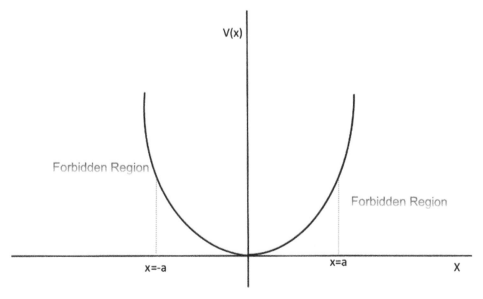

FIGURE 5.1 The graphical form of the potential energy function. It is symmetrical and has parabolic form. At $x = \pm a$, a particle exhibiting harmonic motion will have maximum potential energy and zero kinetic energy, whereas at $x = 0$ the potential energy is zero and maximum kinetic energy. The regions to the left and right of $x = \pm a$ are called forbidden regions.

of the potential energy function of such a system is given as:

$$V(x) = \frac{1}{2}kx^2 \tag{5.1}$$

where $\sqrt{\frac{k}{m}} = \omega$ is the angular frequency of oscillation.

Conceptual Question 1: *Classically, a particle exhibiting harmonic oscillations cannot be found in a forbidden region. Explain why.*

Conceptual Question 2: *Considering Figure 5.1, explain why $x = \pm a$ are called turning points. Does the classical harmonic oscillator spend least or maximum time at $x = 0$?*

The Schrodinger equation for the harmonic oscillator is:

$$-\frac{\hbar^2}{2m}\frac{d^2\psi(x)}{dx^2} + \frac{1}{2}kx^2\psi(x) = E\psi(x) \tag{5.2}$$

By making the substitutions given as follows, the previous equation obtains the following form:

$$x = \sqrt{\frac{\hbar}{m\omega}}y, \quad E = \frac{\hbar\omega}{2}\varepsilon$$

$$\frac{d^2\psi(y)}{dy^2} + (\varepsilon - y^2)\psi(y) = 0 \tag{5.3}$$

Equation (5.3) is not straight forward to solve. Therefore, we need to use an alternative approach to solve this equation. This approach was developed by E. Schrodinger in 1942 and requires factoring the Hamiltonian into two operators, each involving first derivatives as shown in the following equation.

Equation (5.4) can be rewritten as:

$$\left(\frac{d^2}{dy^2} - y^2\right)\psi(y) = -\varepsilon\psi(y) \tag{5.4}$$

The first operator in the previous equation can be described as:

$$\left(\frac{d^2}{dy^2} - y^2\right)\psi = \left[\left(\frac{d}{dy} - y\right)\left(\frac{d}{dy} + y\right) - 1\right]\psi \tag{5.5}$$

By substituting Equation (5.5) into Equation (5.4), the new Schrodinger equation is obtained:

$$\left(\frac{d}{dy} - y\right)\left(\frac{d}{dy} + y\right)\psi_\varepsilon = -(\varepsilon - 1)\psi_\varepsilon \tag{5.6}$$

where ψ_ε is an eigenfunction belonging to eigenvalue ε for the new Schrodinger equation.

By operating with operator $(\frac{d}{dy} + y)$ on the left-hand side of the previous equation,

$$\left(\frac{d}{dy} + y\right)\left(\frac{d}{dy} - y\right)\left(\frac{d}{dy} + y\right)\psi_\epsilon = -(\epsilon - 1)\left(\frac{d}{dy} + y\right)\psi_\epsilon \tag{5.7}$$

Let,

$$\left(\frac{d}{dy} + y\right)\psi_\epsilon = \varphi_\epsilon \tag{5.8}$$

Here, φ_ϵ is the new energy eigenfunction of the Schrodinger equation corresponding to eigenvalue $\epsilon - 2$. Thus, the Schrodinger equation changes to the following form:

$$\left(\frac{d^2}{dy^2} - y^2\right)\varphi_\epsilon = -(\epsilon - 2)\varphi_\epsilon \tag{5.9}$$

Therefore, if ψ_ϵ is the energy eigenfunction of the Schrodinger equation corresponding to eigenvalue ϵ, then φ_ϵ is also the eigenfunction of the Schrodinger equation corresponding to eigenvalue $\epsilon - 2$. This means that for any given solution of the Schrodinger equation we can derive another. Also, if ϵ is the allowed energy eigenvalue, then $\epsilon - 2$ is also the allowed eigenvalue. We can repeat this procedure indefinitely, and we are led to conclude that if ϵ is the eigenvalue, then $\epsilon - 2n$ is also an allowed eigenvalue. However, n cannot be indefinitely big because then the energy of the harmonic oscillator become negative. We know that cannot be true, since the expectation value of energy or average energy must always be positive. This can only be done when the lowest positive value of ϵ is such that $(\frac{d}{dy} + y)\psi_\epsilon = 0$. Multiplication of this equation by the operator $(\frac{d}{dy} - y)$, we get:

$$\left(\frac{d}{dy} - y\right)\left(\frac{d}{dy} + y\right)\psi_\epsilon = 0 \tag{5.10}$$

But from Equation (5.7), we know that the previous equation becomes true only when the lowest energy $\epsilon = 1$. Thus, the only allowed values of ϵ must be such that,

$$1 = \epsilon - 2n \tag{5.11}$$

$$\epsilon = 2n + 1 \tag{5.12}$$

Here, n can only be a positive integer. Therefore, the eigenvalues of E are:

$$E = (2n + 1)\frac{\hbar\omega}{2} = \left(n + \frac{1}{2}\right)\hbar\omega \tag{5.13}$$

In the previous equation, it is clear that the energies of a quantum harmonic oscillator are discrete and have half integral values. The lowest energy of the harmonic oscillator is $1/2\hbar\omega$.

Conceptual Question 3: Explain why the lowest energy value cannot be zero.

The energy eigenfunctions can be obtained by solving the Schrodinger equation for the lowest energy. Instead of ϵ, we will now use "n" in the notation of the wave function:

$$\left(\frac{d}{dy} + y\right)\psi_{n=0} = 0 \tag{5.14}$$

The solution is:

$$\psi_0 = A\, e^{-y^2/2} \tag{5.15}$$

where A is a normalization constant. The previous eigenfunction corresponds to lowest energy state of the harmonic oscillator and is called the ground state energy eigenfunction. All other eigenfunctions can be obtained from the this ground state energy eigenfunction. To do so, we write the Schrodinger equation in the following way:

$$\left(\frac{d}{dy}+y\right)\left(\frac{d}{dy}-y\right)\psi_\epsilon = -(\epsilon+1)\psi_\epsilon \tag{5.16}$$

Multiplication from the left side by $\left(\frac{d}{dy}-y\right)$ yields

$$\left(\frac{d}{dy}-y\right)\left(\frac{d}{dy}+y\right)\left(\frac{d}{dy}-y\right)\psi_\epsilon = -(\epsilon+1)\left(\frac{d}{dy}-y\right)\psi_\epsilon \tag{5.17}$$

$$\left(\frac{d^2}{dy^2}-y^2\right)\left(\frac{d}{dy}-y\right)\psi_\epsilon = -(\epsilon+2)\left(\frac{d}{dy}-y\right)\psi_\epsilon \tag{5.18}$$

We see that the function $\left(\frac{d}{dy}-y\right)\psi_\epsilon = \varphi$ satisfies Schrodinger equation and corresponds to eigenvalue $\epsilon+2$. Therefore, the next higher eigenfunction can be obtained by operating on it with the operator $\left(\frac{d}{dy}-y\right)$.

In this way, we can obtain all the eigenfunctions from ψ_0. Thus, the nth eigenfunction is:

$$\psi_n = A_n(-1)^n\left(\frac{d}{dy}-y\right)^n \psi_0 \tag{5.19}$$

where A_n is normalizing constant. By carrying out these operations, first few eigenfunctions are:

$$\psi_0 = A_0 e^{-y^2/2} \tag{5.20}$$

$$\psi_1 = -A_1 e^{y^2/2}\frac{d}{dy}e^{-y^2} = A_1\left(2ye^{-\frac{y^2}{2}}\right) \tag{5.21}$$

$$\psi_2 = A_2 e^{y^2/2}\frac{d^2}{dy^2}e^{-y^2} = A_2(2y^2-1)e^{-\frac{y^2}{2}} \tag{5.22}$$

$$\psi_n = (-1)^n A_n e^{y^2/2}\frac{d^n}{dy^n}e^{-y^2} \tag{5.23}$$

$$\psi_n = A_n e^{-y^2/2}\frac{d^n}{dy^n}e^{-y^2} \tag{5.24}$$

Therefore, ψ_n is equal to $e^{-y^2/2}$ times some polynomial of nth degree. This polynomial is called Hermite polynomial $h_n(y)$. Thus, one can write:

$$\psi_n = A_n e^{-y^2/2}h_n(y) \tag{5.25}$$

and

$$h_n(y) = \psi_n(y) A_n e^{\frac{y^2}{2}}$$ (5.26)

5.1.1.1 Normalization Constant

The normalization constant can be evaluated using the condition

$$\int_{-\infty}^{\infty} \psi_n \psi_n^* \, dy = (-1)^n |A_n|^2 \int_{-\infty}^{\infty} h_n(y) \frac{d^n}{dy^n} e^{-y^2} \, dy = 1$$ (5.27)

Integrating by parts "n" times, noting that the integrated parts always vanish, and a factor of -1 is introduced each time, this gives us $(-1)^n$ appearing in front of the integral. We obtain:

$$(-1)^{2n} |A_n|^2 \int_{-\infty}^{\infty} e^{-y^2} \frac{d^n h_n(y)}{dy^n} \, dy = 1$$ (5.28)

where $h_n(y)$ is a polynomial of nth degree and can be written as follows:

$$h_n(y) = \sum_{\varphi=0}^{n} B_\varphi y^\varphi$$ (5.29)

The differentiation of the previous polynomial in the integral will wipe out all terms except that involving y^n, and $\frac{d^n y^n}{dy^n} = n!$, therefore

$$\frac{d^n h_n(y)}{dy^n} = n! B_n$$ (5.30)

To evaluate B_n, we note that the coefficient of y^n, we must evaluate $\frac{d^n h_n(y)}{dy^n}$, and according to Rodrigues formula, the Hermite polynomial can be written as $h_n(y) = (-1)^n e^{y^2} \frac{d^n e^{-y^2}}{dy^n}$. Thus, the coefficient of y^n, will be $B_n = 2^n$. Substituting all these coefficients in Equation (5.29), we get:

$$2^n n! |A_n|^2 \int_{-\infty}^{\infty} e^{-y^2} dy = 1, \quad A_n = \frac{1}{(2^n n!)^{1/2}} \frac{1}{(\pi)^{1/4}}$$ (5.31)

Therefore, the normalized wave function in terms of y can be written as:

$$\psi_n(y) = \frac{e^{-y^2/2} h_n(y)}{(\pi)^{1/4} (2^n n!)^{1/2}}$$ (5.32)

Since, $x = \sqrt{\frac{\hbar}{m\omega}} y$, to normalize the previous function over x, we must multiply it by $(\frac{m\omega}{\hbar})^{1/4}$. The normalized wave function of x is:

$$\psi_n(x) = \frac{e^{-(\frac{m\omega}{\hbar})(\frac{x^2}{2})} h_n\left(\sqrt{\frac{m\omega}{\hbar}} x\right)}{(\pi)^{1/4} (2^n n!)^{1/2}} \left(\frac{m\omega}{\hbar}\right)^{1/4}$$ (5.33)

Figure 5.2 describes the wave function and probability density distribution for several energy levels. It also shows that there is a small probability of a particle being found in a forbidden region for all energy levels.

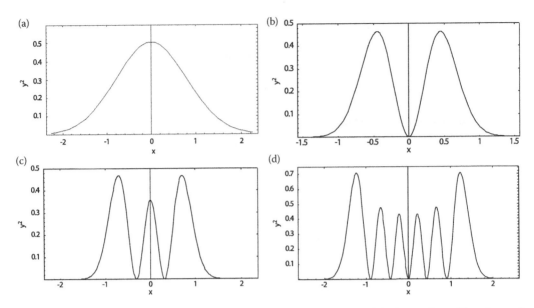

FIGURE 5.2 (a,b) Plots of the probability density (y-axis) and position (x) of the harmonic oscillator in $n = 0$ and $n = 1$ energy states. (c,d) Plots of probability density (y-axis) and position (x) of the harmonic oscillator in $n = 2$ and $n = 5$ energy states.

Let us now discuss the differences between a classical and quantum Harmonic oscillator. Consider a small particle of mass "m" exhibiting harmonic motion about an equilibrium position (Figures 5.3 and 5.4).

Classical Harmonic Oscillator	Quantum Harmonic Oscillator
1. Energy is continuous and can be of any value.	Energy is discrete and only can be of certain values (E_n). All energy levels are equally spaced ($\Delta E = \hbar\omega$). However, for very large "n" (quantum number) energy levels become continuous. $\left(\frac{\Delta E}{E_n} = \frac{\hbar\omega}{(n+\frac{1}{2})\hbar\omega} \rightarrow 0\right)$ as $n \rightarrow \infty$.
2. The particle cannot be in the forbidden region. Its kinetic energy becomes negative in the forbidden region, which violates laws of classical physics.	The particle can be in the forbidden region. It has non-zero probability density in the forbidden region.
3. The particle has maximum kinetic energy at $x = 0$ and, therefore, spends least time at this point.	In the ground state when $n = 0$, also called the lowest energy or zero-point energy, the particle has maximum probability density at $x = 0$.
4. The lowest possible kinetic energy can be zero. Thus, when temperature $T \rightarrow 0K$, K.E $\rightarrow 0$	The lowest possible energy can never be zero. Thus, when temperature $T \rightarrow 0K$, K.E $\neq 0$
5. ???	???

Conceptual Question 4: Consider the ground state of quantum harmonic oscillator as shown in Figure 5.2a. Is the probability density of the quantum harmonic oscillator more or least at $x = 0$? Does this probability density remain the same or change for higher energy states such as $n = 5$? Explain.

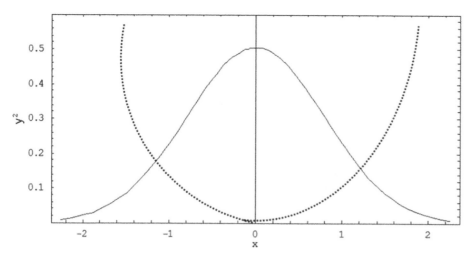

FIGURE 5.3 A comparison of ground state probability density distribution of a quantum harmonic oscillator and a classical oscillator (dotted curve).

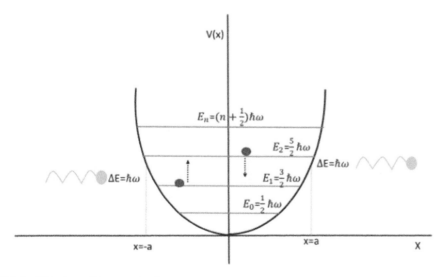

FIGURE 5.4 This figure illustrates that by absorbing a quantum of energy $\Delta E = \hbar\omega$, a quantum harmonic oscillator can transition to the next higher energy level. It also shows that by emitting a quantum of the same energy, the quantum harmonic oscillator can transition to the next lower energy level.

PROBLEMS

5.1 Consider a potential function $V(x)$, and expand it using the Taylor series expansion function. Explain the assumptions you will make to approximate it as a parabolic potential function of a harmonic oscillator. Now consider a spring of spring constant "k" attached to a mass of "m." Show mathematically that classically its average kinetic energy is always equal to average potential energy $(\overline{K} = \overline{V})$.

5.2 Using the wave function of a quantum harmonic oscillator, show that the average kinetic energy is equal to the average potential energy.

5.3 Consider a pendulum exhibiting harmonic motion. It is simply a mass "m" attached to string that moves about an angle "θ." What will happen to its angle "θ," if it is a quantum harmonic oscillator? Show mathematically the quantization of the angle.

5.4 Consider an electron exhibiting simple harmonic motion in the ground state. What should be "k" so that it oscillates with 10 nHz frequency? What will be its maximum displacement "a"?

5.5 Compute $\langle x \rangle$, $\langle x^2 \rangle$, and $\langle p \rangle$ for the ground state wave function?

5.6 A particle is in linear superposition state $\Psi(x, 0) = A(2\psi_0(x) + \psi_2(x))$. Find the normalization constant "A." Construct a time-dependent linear superposition state. Sketch the probability densities at $t = 0$ and "t."

5.1.2 ALGEBRAIC METHOD

In the last section, we learned the analytic method for solving the Schrodinger equation for the harmonic oscillator. In this section, the algebraic method is described, which was founded by P.A.M Dirac.

The Hamiltonian operator of the harmonic oscillator is:

$$\hat{H} = \hat{T} + \hat{V} = \frac{\hat{p}^2}{2m} + \frac{1}{2}m\omega^2\hat{x}^2 \tag{5.34}$$

where $\hat{p}$ and $\hat{x}$ are operators and not numbers, and $\hat{p} = \frac{\hbar}{i}\frac{d}{dx}$ is a momentum operator. Is there a way to write $\hat{p}$ and $\hat{x}$ so that we can describe them as a single operator? Yes! Let us see how.

Let us define new operators as:

$$\hat{a}_+ = \frac{1}{\sqrt{2\hbar\omega m}}(\hat{x} - i\hat{p}) \tag{5.35}$$

$$\hat{a}_- = \frac{1}{\sqrt{2\hbar\omega m}}(\hat{x} + i\hat{p}) \tag{5.36}$$

Since the previous are operators, their order matters. And,

$$\hat{a}_+\hat{a}_- \neq \hat{a}_-\hat{a}_+ \tag{5.37}$$

$$\hat{a}_+\hat{a}_- = \frac{1}{2\hbar m\omega}[\hat{p}^2 + (m\omega\hat{x})^2 + im\omega(\hat{x}\hat{p} - \hat{p}\hat{x})] \tag{5.38}$$

$$\hat{a}_-\hat{a}_+ = \frac{1}{2\hbar m\omega}[\hat{p}^2 + (m\omega\hat{x})^2 - im\omega(\hat{x}\hat{p} - \hat{p}\hat{x})] \tag{5.39}$$

Subtracting Equation (5.38) from Equation (5.39), we get

$$\hat{a}_-\hat{a}_+ - \hat{a}_+\hat{a}_- = \frac{-im\omega(\hat{x}\hat{p} - \hat{p}\hat{x})}{2\hbar m\omega} - \frac{im\omega(\hat{x}\hat{p} - \hat{p}\hat{x})}{2\hbar m\omega} \tag{5.40}$$

$$\hat{a}_-\hat{a}_+ - \hat{a}_+\hat{a}_- = \frac{-i(\hat{x}\hat{p} - \hat{p}\hat{x})}{\hbar} \tag{5.41}$$

$$\hat{x}\hat{p} - \hat{p}\hat{x} = [\hat{x}, \hat{p}] \tag{5.42}$$

In quantum mechanics, the previous relationship of operators is a momentum and position operators commutator relationship. This relationship is simply, $\hat{x}\hat{p} - \hat{p}\hat{x} = [\hat{x}, \hat{p}] = i\hbar$, and substituting this value in Equation (5.41), we get:

$$\hat{a}_- \hat{a}_+ - \hat{a}_+ \hat{a}_- = 1 \tag{5.43}$$

$$\hat{a}_- \hat{a}_+ + \hat{a}_+ \hat{a}_- = \frac{1}{\hbar m \omega} [\hat{p}^2 + (m \omega \hat{x})^2] = \frac{2\hat{H}}{\hbar m \omega} \tag{5.44}$$

Substituting Equation (5.43) in Equation (5.44), we get:

$$\hbar \omega \left[\hat{a}_+ \hat{a}_- + \frac{1}{2} \right] = \hat{H} \tag{5.45}$$

The Schrodinger equation can now be written as:

$$\hat{H}\psi = \hbar \omega \left[\hat{a}_+ \hat{a}_- + \frac{1}{2} \right] \psi \tag{5.46}$$

So far, we have used the tricks to describe the Hamiltonian, and we have simply written the Schrodinger equation in a mathematically different form, but we have yet not solved the Schrodinger equation. Let us see how we can solve the Schrodinger using the previous mathematical form. To do this, we need to see the effect of operators $\hat{a}_+$ and $\hat{a}_-$ on the energy eigenfunction ψ.

$$\hat{H}(\hat{a}_+ \psi) = \hbar \omega \left[\hat{a}_+ \hat{a}_- \hat{a}_+ + \frac{\hat{a}_+}{2} \right] \psi \tag{5.47}$$

Since,

$$\hat{a}_+ \hat{a}_- \hat{a}_+ + \frac{\hat{a}_+}{2} = \frac{\hat{a}_+}{2} \left[\hat{a}_+ \hat{a}_- + 1 + \frac{1}{2} \right]$$

Thus,

$$\hat{H}(\hat{a}_+ \psi) = \frac{\hat{a}_+}{2} \left[\hbar \omega \left(\hat{a}_+ \hat{a}_- + 1 + \frac{1}{2} \right) \right] \psi \tag{5.48}$$

If

$$\hat{H}\psi = E\psi = \hbar \omega \left[\hat{a}_+ \hat{a}_- + \frac{1}{2} \right] \psi$$

Then,

$$\hat{H}(\hat{a}_+ \psi) = \hat{a}_+ [E + \hbar \omega] \psi \tag{5.49}$$

This shows that the action of operator $\hat{a}_+$ raises the energy level by 1. If you continue operating this operator, then it keeps increasing the energy level. Similarly, the effect of $\hat{a}_-$ on the energy eigenfunction will be to lower the energy by 1 as shown by:

$$\hat{H}(\hat{a}_- \psi) = \hat{a}_- [E - \hbar \omega] \psi \tag{5.50}$$

Let us call $\hat{a}_+$ a raising operator and $\hat{a}_-$ a lowering operator, but if we apply the lowering operator repeatedly, how much lower it can go? We know that energy cannot be negative. Therefore, if we repeatedly apply the lowering operator, we will reach a level where applying this operator will give zero energy. Thus,

$$\hat{a}_-\psi_0 = 0 \tag{5.51}$$

We can use the above equation to determine ψ_0:

$$\frac{1}{\sqrt{2\hbar\omega m}}\left(m\omega\hat{x} + \hbar\frac{d}{dx}\right)\psi_0 = 0 \tag{5.52}$$

The previous equation is a first-order ordinary differential equation whose solution is trivial and given as:

$$\psi_0 = Ae^{\frac{-m\omega}{2\hbar}x^2} \tag{5.53}$$

The previous wave function is the energy eigenfunction of the lowest possible energy state. This lowest energy eigenvalue can be obtained by operating Hamiltonian operator over this eigenfunction.

$$\hat{H}\psi_0 = \hbar\omega\left[\hat{a}_+\hat{a}_- + \frac{1}{2}\right]\psi_0 \tag{5.54}$$

Noting that $\hat{a}_-\psi_0 = 0$, we get the following:

$$\hat{H}\psi_0 = \frac{\hbar\omega}{2}\psi_0 \tag{5.55}$$

Therefore, the lowest energy of the harmonic oscillator is:

$$E_0 = \frac{\hbar\omega}{2} \tag{5.56}$$

To generate wave functions of other higher states, we operate raising operators repeatedly. For example, to generate nth wave function or energy eigenfunction, we operate raising operator "n" times on the ground state as shown in the following equation:

$$\psi_n = A_n(\hat{a}_+)^n\psi_0 \tag{5.57}$$

A_n is a normalization constant, and the corresponding energy eigenvalue is:

$$E_n = \left(n + \frac{1}{2}\right)\hbar\omega \tag{5.58}$$

EXAMPLE 5.1

Find the normalization constant for the ground state energy eigenfunction. Construct the energy eigenfunction for $n = 1$ state using Equation (5.57).

$$\psi_0 = Ae^{\frac{-m\omega}{2\hbar}x^2}$$

$$1 = |A|^2\int_{-\infty}^{\infty} e^{\frac{-m\omega}{\hbar}x^2}\,dx = |A|^2\sqrt{\frac{\pi\hbar}{m\omega}}$$

$$A = \left(\frac{m\omega}{\pi\hbar}\right)^{1/4} \tag{5.59}$$

$$\psi_0 = \left(\frac{m\omega}{\pi\hbar}\right)^{1/4} e^{\frac{-m\omega}{2\hbar}x^2}$$

For $n = 2$ energy eigenstate, the eigenfunction is:

$$\psi_1(x) = A_1(\hat{a}_+)^1\psi_0 = A_1\left(\frac{m\omega}{\pi\hbar}\right)^{1/4}\sqrt{\frac{2m\omega}{\hbar}}xe^{-\frac{m\omega}{2\hbar}x^2} \tag{5.60}$$

Find A_1 using the normalization conditions, and you will find that it is simply 1. Thus,

$$\psi_1(x) = A_1(\hat{a}_+)^1\psi_0 = \left(\frac{m\omega}{\pi\hbar}\right)^{1/4}\sqrt{\frac{2m\omega}{\hbar}}xe^{-\frac{m\omega}{2\hbar}x^2} \tag{5.61}$$

5.1.2.1 Normalization Constant

The normalization constant in Equation (5.57) is not derived yet, without which our energy eigenfunction is incomplete. Here, for simplicity, we will describe the form of the normalization constant without derivation of it. Thus,

$$A_n = \frac{1}{\sqrt{n!}}$$

and

$$\psi_n = \frac{1}{\sqrt{n!}}(\hat{a}_+)^n\psi_0 \tag{5.62}$$

$$\psi_n = \frac{1}{\sqrt{n!}}(\hat{a}_+)^n\psi_0 \tag{5.63}$$

$$\psi_{n+1} = \frac{1}{\sqrt{n+1!}}(\hat{a}_+)^{n+1}\psi_0 \tag{5.64}$$

Thus

$$\hat{a}_+\psi_n = \sqrt{n+1}\,\psi_{n+1} \tag{5.65}$$

Similarly,

$$\hat{a}_-\psi_n = \sqrt{n}\,\psi_{n-1} \tag{5.66}$$

You can see that using above relationship we can get:

$$\hat{a}_+\psi_0 = \sqrt{1}\,\psi_1 \tag{5.67}$$

$$\hat{a}_+\psi_1 = \sqrt{2}\,\psi_2 \tag{5.68}$$

Thus,

$$\psi_2 = \frac{\hat{a}_+\psi_1}{\sqrt{2}} \tag{5.69}$$

You can find all other energy eigenfunctions using above two relationships.

EXAMPLE 5.2

Construct the energy eigenfunction for the energy state $n = 2$ using Equation (5.69),

$$\psi_2 = \frac{\hat{a}_+ \psi_1}{\sqrt{2}} = \frac{1}{\sqrt{2}} \frac{1}{\sqrt{2\hbar\omega m}} \left(-\hbar \frac{d}{dx} + m\omega x\right) \left(\frac{m\omega}{\pi\hbar}\right)^{1/4} \sqrt{\frac{2m\omega}{\hbar}} x e^{\frac{-m\omega}{2\hbar}x^2} \qquad (5.70)$$

$$\psi_2 = \frac{\hat{a}_+ \psi_1}{\sqrt{2}} = \frac{1}{\sqrt{2}} \frac{1}{\sqrt{2\hbar\omega m}} \left(\frac{m\omega}{\pi\hbar}\right)^{1/4} \sqrt{\frac{2m\omega}{\hbar}} (-\hbar + 2m\omega x^2) e^{\frac{-m\omega}{2\hbar}x^2} \qquad (5.71)$$

$$\psi_2 = \frac{\hat{a}_+ \psi_1}{\sqrt{2}} = \frac{1}{\sqrt{2}} \left(\frac{m\omega}{\pi\hbar}\right)^{1/4} \left(-1 + \frac{2m\omega x^2}{\hbar}\right) e^{\frac{-m\omega}{2\hbar}x^2} \qquad (5.72)$$

PROBLEMS

5.7 By using the previous method, derive the normalization constant as $A_n = \frac{1}{\sqrt{n!}}$.

5.8 Construct the energy eigenfunction for the energy state $n = 5$ and plot the probability density for such an eigenfunction.

5.2 THE SCHRODINGER EQUATION IN THREE DIMENSIONS

Up to now, we have dealt with 1-dimensional systems whose Schrodinger equations we tried to solve. However, most real systems are 3-dimensional. In this chapter, we will develop approaches for solving 3-dimensional Schrodinger equation. We will also closely study the interesting quantum mechanical properties that can be only determined by treating a system as a 3-dimensional quantum mechanical system.

5.2.1 THE SCHRODINGER EQUATION IN CARTESIAN COORDINATES

The 3-dimensional Schrodinger equation can be simply written as:

$$i\hbar \frac{\partial \Psi(x, y, z, t)}{\partial t} = \hat{H} \, \Psi(x, y, z, t) \qquad (5.73)$$

where

$$\hat{H} = -\frac{\hbar^2}{2m} \nabla^2 + V \quad \text{and} \quad \nabla^2 = \frac{\partial^2}{\partial x^2} + \frac{\partial^2}{\partial y^2} + \frac{\partial^2}{\partial z^2}$$

is a Laplacian in Cartesian coordinates.

The potential energy V and wave function Ψ are functions of coordinates x, y, z and time "t." If "r" is vector in a 3-dimensional space that describes the location of a particle, the probability of finding the particle in an infinitesimal volume is

$$|\Psi|^2 d^3 r \qquad (5.74)$$

and the normalization condition is

$$\int |\Psi|^2 d^3 r = 1 \qquad (5.75)$$

with the integral over all the space. If the potential is time independent, then the solution of the Schrodinger equation will give a complete set of energy eigenfunctions which are stationary states, shown as:

$$\Psi_n(r, t) = \psi_n(r)e^{-iE_n t} \tag{5.76}$$

Using these energy eigenfunctions, the most general form of the wave function for a time-independent potential is written as:

$$\Psi(r, t) = \sum b_n \psi_n(r)e^{-iE_n t} \tag{5.77}$$

where b_n are the coefficients that can be determined by using initial wave function method as described in Chapters 3 and 4.

EXAMPLE 5.3

Consider a 3-dimensional potential well, whose potential is shown as:

$$V(x, y, z) = \begin{cases} 0, & 0 \leq x, y, z \leq L \quad \text{(Region II)} \\ \infty, & \text{otherwise} \quad\quad\;\; \text{(Region I and III)} \end{cases} \tag{5.78}$$

The time-independent Schrodinger equation for such a system is

$$-\frac{\hbar^2}{2m}\nabla^2\Psi(x, y, z) = E\,\Psi(x, y, z) \tag{5.79}$$

The previous equation can be solved by using a method of *separation of variables*. Using this method, the wave function can be described as:

$$\Psi(x, y, z) = X(x)Y(y)Z(z) \tag{5.80}$$

Substituting the wave function in Equation (5.80) into Equation (5.79), we get the following equation:

$$-\frac{\hbar^2}{2m}\left[\frac{1}{X}\frac{d^2X}{dx^2} + \frac{1}{Y}\frac{d^2Y}{dy^2} + \frac{1}{Z}\frac{d^2Z}{dz^2}\right] = E \tag{5.81}$$

Let us define,

$$k^2 = \frac{2mE}{\hbar^2} = k_x^2 + k_y^2 + k_z^2 \tag{5.82}$$

Using this relation, Equation (5.81) can be rewritten as:

$$\left[\frac{1}{X}\frac{d^2X}{dx^2} + k_x^2\right] + \left[\frac{1}{Y}\frac{d^2Y}{dy^2} + k_y^2\right] + \left[\frac{1}{Z}\frac{d^2Z}{dz^2} + k_z^2\right] = 0 \tag{5.83}$$

Thus, each term must be equated to zero and can be written as:

$$\frac{d^2X}{dx^2} = -k_x^2 X(x), \quad \frac{d^2Y}{dy^2} = -k_y^2 Y(y), \quad \frac{d^2Z}{dz^2} = -k_z^2 Z(z) \tag{5.84}$$

The solution of each independent equation using boundary conditions is:

$$\Psi_{n_x,n_y,n_z}(x, y, z) = \left(\frac{2}{L}\right)^{3/2} \sin\left(\frac{n_x \pi}{L}\right)\sin\left(\frac{n_y \pi}{L}\right)\sin\left(\frac{n_z \pi}{L}\right) \tag{5.85}$$

where n_x, n_y, $n_z = 1, 2, 3\dots$ positive integers, also called occupancy quantum numbers.

The total energy E of the system is:

$$E_{n_x,n_y,n_z} = \frac{\pi^2 \hbar^2}{2mL^2}\left(n_x^2 + n_y^2 + n_z^2\right) \tag{5.86}$$

5.2.1.1 What Is the New Quantum Property of the 3-Dimensional Infinite Square Well?

The new property of the system comprising of a particle in a 3-dimensional infinite square well is the increase in the number of available energy levels and degenerate states associated with each level. Each quantum number n_x, n_y, n_z corresponding to each dimension contributes towards these two factors. Therefore, a particle in such a well will have more available energy states and degenerate energy states than a particle in a 1-dimensional infinite square well. The schematic that follows (Figures 5.5 and 5.6) describes the 3-dimensional infinite square well.

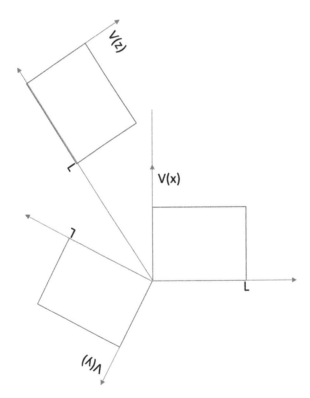

FIGURE 5.5 Visualization of 3-dimensional infinite square well potential.

FIGURE 5.6 Degenerate energy levels of 3-dimensional infinite square well potential.

PROBLEMS

5.9 Using the approach previously discussed, determine a 3-dimensional wave function of a free particle.

5.10 Derive the wave function given in Equation (5.85) by applying the boundary conditions for the 3-dimensional infinite square well (Figures 5.5 and 5.6).

5.2.2 THE SCHRODINGER EQUATION IN SPHERICAL COORDINATES

You may be wondering why we need to describe the Schrodinger equation in spherical coordinates. The main reason for doing so is that often the potential energy of quantum particles is a function of distance r, from some center of symmetry, which may be the center of an atom. Such situations require circular, cylindrical and spherical symmetry. Thus, it is more convenient to express the Schrodinger equation in spherical coordinates. In Figure 5.7, spherical coordinates radius r, polar angle θ and azimuthal angle ϕ are shown.

In spherical coordinates (r, θ, ϕ) the form of Schrodinger equation remains the same. However, the Laplacian can now be written as:

$$\nabla^2 = \frac{1}{r^2}\frac{\partial}{\partial r}\left(r^2\frac{\partial}{\partial r}\right) + \frac{1}{r^2\sin\theta}\frac{\partial}{\partial\theta}\left(\sin\theta\frac{\partial}{\partial\theta}\right) + \frac{1}{r^2\sin^2\theta}\left(\frac{\partial^2}{\partial\phi^2}\right) \tag{5.87}$$

In spherical coordinates, the time-independent Schrodinger equation is:

$$\left[-\frac{\hbar^2}{2m}\left\{\frac{1}{r^2}\frac{\partial}{\partial r}\left(r^2\frac{\partial}{\partial r}\right) + \frac{1}{r^2\sin\theta}\frac{\partial}{\partial\theta}\left(\sin\theta\frac{\partial}{\partial\theta}\right) + \frac{1}{r^2\sin^2\theta}\left(\frac{\partial^2}{\partial\phi^2}\right)\right\} + V\right]\Psi = E\,\Psi \tag{5.88}$$

Using the method of separation of variables, we can write the wave function as:

$$\Psi(r, \theta, \phi) = R(r)Y(\theta, \phi) \tag{5.89}$$

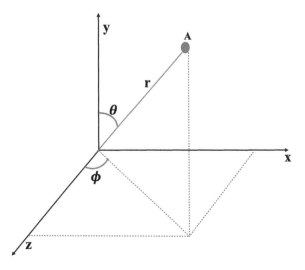

FIGURE 5.7 Spherical coordinates of a point particle at a distance "r" from the center of symmetry.

Substituting Equation (5.89) into Equation (5.88), we get:

$$-\frac{\hbar^2}{2m}\left[\frac{Y}{r^2}\frac{d}{dr}\left(r^2\frac{dR}{dr}\right)+\frac{R}{r^2\sin\theta}\frac{\partial}{\partial\theta}\left(\sin\theta\frac{\partial Y}{\partial\theta}\right)+\frac{R}{r^2\sin^2\theta}\left(\frac{\partial^2 Y}{\partial\phi^2}\right)\right]+VRY=E\,RY \qquad (5.90)$$

Dividing by RY and multiplying by $-2mr^2/\hbar^2$, the following equation is obtained:

$$\left\{\frac{1}{R}\frac{d}{dr}\left(r^2\frac{dR}{dr}\right)-\frac{2mr^2}{\hbar^2}[V(r)-E]\right\}+\frac{1}{Y}\left\{\frac{1}{\sin\theta}\frac{\partial}{\partial\theta}\left(\sin\theta\frac{\partial Y}{\partial\theta}\right)+\frac{1}{\sin^2\theta}\left(\frac{\partial^2 Y}{\partial\phi^2}\right)\right\}=0 \qquad (5.91)$$

The previous equation has two terms. The first term entirely depends on R, and the second term depends on coordinates θ, ϕ. Thus, each term must be equal to some constant, and it breaks down into the following two equations:

$$\frac{1}{R}\frac{d}{dr}\left(r^2\frac{dR}{dr}\right)-\frac{2mr^2}{\hbar^2}[V(r)-E]=-C \qquad (5.92)$$

$$\frac{1}{Y}\left[\frac{1}{\sin\theta}\frac{\partial}{\partial\theta}\left(\sin\theta\frac{\partial Y}{\partial\theta}\right)+\frac{1}{\sin^2\theta}\left(\frac{\partial^2 Y}{\partial\phi^2}\right)\right]=C \qquad (5.93)$$

Equation (5.92) is called the *radial equation*, and Equation (5.93) is called the *angular equation*. Now, we need to obtain solutions of these equations and then combine them to form a solution of the Schrodinger equation.

Conceptual Question 5: *Explain in your own words the advantages of using spherical coordinates over the Cartesian coordinate system for describing a 3-dimensional quantum mechanical system that is under the influence of a potential V(r).*

5.2.3 THE ANGULAR EQUATION

The angular equation is associated with the total angular momentum of the particle. Therefore, the solution of this equation gives the eigenvalues of the total angular momentum. The angular momentum operators in spherical coordinates are described as:

$$\hat{L}_z=\frac{\hbar}{i}\frac{\partial}{\partial\phi}$$

$$\hat{L}_x=-\frac{\hbar}{i}\left(\sin\phi\frac{\partial}{\partial\theta}+\cot\theta\cos\phi\frac{\partial}{\partial\phi}\right) \qquad (5.94)$$

$$\hat{L}_y=\frac{\hbar}{i}\left(\cos\phi\frac{\partial}{\partial\theta}-\cot\theta\sin\phi\frac{\partial}{\partial\phi}\right)$$

In spherical coordinates, using the previous operators, the square of the total angular momentum is computed as:

$$\hat{L}^2=\hat{L}_x^2+\hat{L}_y^2+\hat{L}_z^2=-\hbar^2\left[\frac{1}{\sin\theta}\frac{\partial}{\partial\theta}\left(\sin\theta\frac{\partial}{\partial\theta}\right)+\frac{1}{\sin^2\theta}\left(\frac{\partial^2}{\partial\phi^2}\right)\right] \qquad (5.95)$$

Thus,

$$\hat{L}^2 f=cf \qquad (5.96)$$

where $c =$ the eigenvalue of the operator $\hat{L}^2$, and $f = s$ an eigenfunction. It is customary to write the eigenvalue of this operator as $\hbar^2 l(l+1)$, where l is a positive integer and is called the *total angular momentum quantum number*. It is also called the *azimuthal quantum number*. The word azimuth refers to the spherical angle. Thus,

$$\hat{L}^2 f = \hbar^2 l(l+1)f \tag{5.97}$$

In terms of the previous operator, the angular equation can be written as:

$$\left[\frac{1}{\sin\theta}\frac{\partial}{\partial\theta}\left(\sin\theta\frac{\partial Y}{\partial\theta}\right) + \frac{1}{\sin^2\theta}\left(\frac{\partial^2 Y}{\partial\phi^2}\right) \right] = CY = \frac{-\hat{L}^2 Y}{\hbar^2} = -l(l+1)Y \tag{5.98}$$

Thus,

$$C = -l(l+1) \tag{5.99}$$

If we now go back to the Schrodinger equation and write the Schrodinger equation in terms of the angular momentum operator, we get the following equation:

$$\left[-\frac{\hbar^2}{2m}\left\{ \frac{1}{r^2}\frac{\partial}{\partial r}\left(r^2\frac{\partial}{\partial r}\right) \right\} + \frac{\hat{L}^2}{2mr^2} + V \right]\Psi = E\,\Psi \tag{5.100}$$

The Hamiltonian operator is then

$$\hat{H} = -\frac{\hbar^2}{2m}\left\{ \frac{1}{r^2}\frac{\partial}{\partial r}\left(r^2\frac{\partial}{\partial r}\right) \right\} + \frac{\hat{L}^2}{2mr^2} + V(r) \tag{5.101}$$

since $V = V(r)$ is not a function of coordinates (θ, ϕ), $\hat{L}^2$ and $\hat{L}$ commutes with $\hat{H}$, which means that these operators are constants of motion. Therefore, the expectation values of these operators will not change with time, and they are called quantum-mechanical constants. The eigenfunctions of these operators will also be the eigenfunctions of the Hamiltonian, and as was shown in Equation (5.89), we need to now find these eigenfunctions by solving the angular equation.

Let us solve this equation by again using the method of separation of variables; therefore,

$$Y(\theta, \phi) = \Theta(\theta)\Phi(\phi) \tag{5.102}$$

By substituting the previous function in the angular equation (Equation 5.98), we get following equations:

$$\left\{ \frac{1}{\Theta}\left[\sin\theta\frac{d}{d\theta}\left(\sin\theta\frac{d\Theta}{d\theta}\right) \right] + l(l+1)\sin^2\theta \right\} + \frac{1}{\Phi}\left(\frac{d^2\Phi}{d\phi^2}\right) = 0 \tag{5.103}$$

The first term is a function of θ, and the second term is a function of ϕ only. Thus, each must be equal to some same constant.

The previous two equations are:

$$\frac{1}{\Theta}\left[\sin\theta\frac{d}{d\theta}\left(\sin\theta\frac{d\Theta}{d\theta}\right) \right] + l(l+1)\sin^2\theta = m^2 \tag{5.104}$$

$$\frac{1}{\Phi}\left(\frac{d^2\Phi}{d\phi^2}\right) = -m^2 \tag{5.105}$$

By solving this equation, we get:

$$\Phi(\phi) = e^{im\phi} \tag{5.106}$$

There are two solutions ($e^{im\phi}$ and $e^{-im\phi}$), but we will cover the other solution by allowing m to have both positive and negative values. Since,

$$\Phi(\phi) = \Phi(\phi + 2\pi) \tag{5.107}$$

thus, m must be an integer and is called the *magnetic quantum number*

$$m = 0, \pm 1, \pm 2, \pm 3, \pm 4, \ldots \tag{5.108}$$

With this, the remaining angular equation becomes:

$$\left[\sin\theta \frac{d}{d\theta}\left(\sin\theta \frac{d\Theta}{d\theta} \right) \right] + [l(l+1)\sin^2\theta - m^2]\Theta = 0 \tag{5.109}$$

The solution of the previous equation is not trivial and is shown as:

$$\Theta(\theta) = AP_l^m(\cos\theta) \tag{5.110}$$

where P_l^m is the associated Legendre function defined as:

$$P_l^m(x) = (1 - x^2)^{\frac{|m|}{2}}\left(\frac{d}{dx} \right)^{|m|} P_l(x) \tag{5.111}$$

and $P_l(x)$ is the lth Legendre polynomial defined by Rodrigues formula:

$$P_l(x) = \frac{1}{2^l l!}\left(\frac{d}{dx} \right)^l (x^2 - 1)^l \tag{5.112}$$

Notice that in this equation, l must be a non-negative number. Otherwise, the mathematical form of the function does not make sense.

The normalized angular wave functions are called *spherical harmonics*,

$$Y_l^m = b\sqrt{\frac{(2l+1)(l-|m|)!}{4\pi\,(l+|m|)!}}\,e^{im\phi}P_l^m(\cos\theta) \tag{5.113}$$

TABLE 5.1

The Mathematical Form of (a) Legendre Polynomials, $P_l(x)$, and (b) Associated Legendre Functions, $P_l^m(\cos\theta)$

$P_l(x)$		$P_l^m(\cos\theta)$	
$P_0(x)$	1	P_0^0	1
$P_1(x)$	x	P_1^1	$\sin\theta$
$P_2(x)$	$\frac{1}{2}(3x^2 - 1)$	P_1^0	$\cos\theta$
$P_3(x)$	$\frac{1}{2}(5x^3 - 3x)$	P_2^2	$3\sin^2\theta$
$P_4(x)$	$\frac{1}{8}(35x^4 - 30x^2 + 3)$	P_2^1	$3\sin^2\theta\cos\theta$
		P_2^0	$\frac{1}{2}(3\cos^2\theta - 1)$

TABLE 5.2

The First Few Spherical Harmonics, $Y_l^m(\theta, \phi)$

Y_0^0	$\left(\frac{1}{4\pi}\right)^{1/2}$
Y_1^0	$\left(\frac{3}{4\pi}\right)^{1/2}\cos\theta$
$Y_1^{\pm 1}$	$\mp\left(\frac{3}{8\pi}\right)^{1/2}\sin\theta e^{\pm i\phi}$
Y_2^0	$\left(\frac{5}{16\pi}\right)^{1/2}(3\cos^2\theta - 1)$
$Y_2^{\pm 1}$	$\left(\frac{15}{8\pi}\right)^{1/2}\sin\theta\cos\theta e^{\pm i\phi}$

---

where $b = (-1)^m$ for $m \geq 0$ and $b = 1$ for $m \leq 0$, and these functions are orthogonal. The properties of these functions (Tables 5.1 and 5.2) will become critical when we discuss hydrogen atom. Until then, we will just see them as mathematical functions without going into details. Interestingly, Equation (5.113) does not directly depend on the form of the potential function but rather on the symmetry of spherical coordinate system.

5.2.4 THE RADIAL EQUATION

The equation that follows depends on coordinate r and potential function $V(r)$. To solve this equation, we need to know the form of the potential function. In this section, we will not solve this equation. Instead, we will closely look at its form.

$$\frac{1}{R}\frac{d}{dr}\left(r^2\frac{dR}{dr}\right) - \frac{2mr^2}{\hbar^2}[V(r) - E] = l(l+1)$$

$$\frac{1}{R}\frac{d}{dr}\left(r^2\frac{dR}{dr}\right) - \frac{2mr^2}{\hbar^2}\left[V(r) + \frac{\hbar^2}{2mr^2}l(l+1) - E\right] = 0 \tag{5.114}$$

This equation can be further simplified by substituting the following function:

$$u(r) = rR(r) \tag{5.115}$$

and the equation becomes:

$$-\frac{\hbar^2}{2m}\frac{d^2u}{dr^2} + \left[V(r) + \frac{\hbar^2}{2mr^2}l(l+1)\right]u = Eu \tag{5.116}$$

This equation is identical to a 1-dimensional Schrodinger equation except for the second term. The second term is called effective potential:

$$V_{\mathit{eff}} = V(r) + \frac{\hbar^2}{2mr^2}l(l+1) \tag{5.117}$$

The second term in the previous equation is due to angular momentum, and it acts as a repulsive potential. This repulsive potential is similar to centrifugal force in that it is responsible for keeping the particle from having angular momentum away from the origin. For particles with $l = 0$, (zero angular momentum), this potential becomes zero. With all this understanding, we are prepared to tackle the problem of hydrogen atom in the next section.

5.3 THE HYDROGEN ATOM

In this section of the chapter, we will solve the three-dimensional Schrodinger equation for the hydrogen atom. With our understanding from the previous sections, solving such an equation becomes easy. For the hydrogen atom, the potential energy function is dependent only on radial distance and is independent of angular variables. Therefore, the solution of the angular equation that we solved in the last section will remain same. Here, we just need to solve the radial equation using the potential energy function for the hydrogen atom. After solving the radial equation, we will combine the solutions of these two equations to determine the wave function of the hydrogen atom.

The hydrogen atom consists of a positively charged proton surrounded by an electron that orbits around it. The charges of the proton and the electron are exactly equal and opposite, with the proton denoted as $+e$ or simply e and the electron as $-e$. Considering that the electron having mass m_e and the proton having mas "m_p" are located at distances, r_1 and r_2 from the origin (Figure 5.8), the forces exerted on these particles are derived from potential energy, which is a function of the relative distance between them $r_1 - r_2$ and is $V(r_1 - r_2)$. According to classical mechanics, such a system is described by the following Hamiltonian:

$$H = \frac{1}{2}m_e\dot{r}_1^2 + \frac{1}{2}m_p\dot{r}_2^2 + V(r_1 - r_2) \tag{5.118}$$

The first two terms are the kinetic energies of the electron and the proton, and the last term is the relative potential energy. The study of the motion of this system can be simplified by introducing the center of mass of the system. The center of mass is described as a point in space where the whole mass of the system is located (Figure 5.8). Thus, the problem simplifies to studying the center of the system, which is responsible for the entire motion of the system. The coordinate of center of mass is given as:

$$r_{cm} = \frac{m_e r_1 + m_p r_2}{m_e + m_p} \tag{5.119}$$

$$r = r_1 - r_2 \tag{5.120}$$

Using Equations (5.119) and (5.120), following relationships can be derived:

$$r_1 = r_{cm} + \frac{m_p r}{m_e + m_p} \tag{5.121}$$

$$r_2 = r_{cm} + \frac{m_e r}{m_e + m_p} \tag{5.122}$$

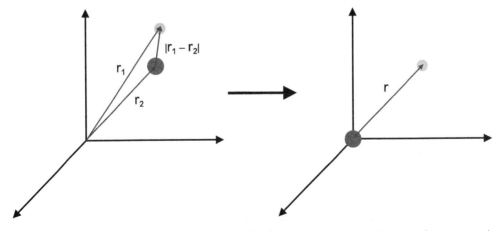

FIGURE 5.8 Illustration of spatial position of a proton and an electron in space and center of mass approximation where a proton occupies the same position as the center of mass.

Substituting Equations (5.121) and (5.122) into Equation (5.118), we get the following Hamiltonian:

$$H = \frac{1}{2}M\dot{r}_{cm}^2 + \frac{1}{2}\mu\dot{r}^2 + \frac{2m_pm_e\dot{r}\dot{r}_{cm}}{(m_e+m_p)^2} + V(r) \tag{5.123}$$

where $M = m_e + m_p =$ the total mass of the system, and $\mu = \frac{m_pm_e}{m_e+m_p}$ the reduced mass of the system. Since all the forces are internal, there are no external forces acting on the center of mass, $\dot{r}_{cm} = 0$, and Equation (5.123) reduces to the following form:

$$H = \frac{1}{2}\mu\dot{r}^2 + V(r) \tag{5.124}$$

Further, the mass of the proton is 1.7×10^{-27} kg, and the mass of the electron is 0.91×10^{-30} kg. It is fine to assume that $m_p \gg m_e$, and $\mu \cong m_e = m$. This also means that the center of mass position is the same as the location of the proton, and relative particle is the electron. Thus, with these assumptions, we have reduced the two-particle system to a system comprised of a single particle placed in the potential $V(r)$. The form of the potential is the electrostatic potential energy given by the Coulomb's law as:

$$V(r) = -\frac{e^2}{4\pi\varepsilon_0}\frac{1}{r} \tag{5.125}$$

Conceptual Question 6: *Why is the center of mass of an electron-proton system and a proton located at the same position?*

5.3.1 THE RADIAL EQUATION FOR THE HYDROGEN ATOM

The radial equation for the hydrogen atom can now be written as:

$$-\frac{\hbar^2}{2m}\frac{d^2u}{dr^2} + \left[-\frac{e^2}{4\pi\varepsilon_0}\frac{1}{r} + \frac{\hbar^2}{2mr^2}l(l+1)\right]u = Eu \tag{5.126}$$

To simplify writing the previous equation, we denote the following coefficients:

$$k^2 = \frac{-2mE}{\hbar^2}, \quad \rho = k\,r \quad \text{and } \rho_0 = \frac{me^2}{2\pi\hbar^2k}$$

Now, substitute these coefficients in Equation (5.126), the equation scales to the following form:

$$\frac{d^2u}{d\rho^2} + \left[-1 + \frac{\rho_0}{\rho} - \frac{l(l+1)}{\rho^2}\right]u = 0 \tag{5.127}$$

5.3.1.1 Asymptotic Behavior

As $\rho \to \infty$, the second and third terms in the bracket reduce to zero, and the equation reduces to following form:

$$\frac{d^2u}{d\rho^2} - u = 0 \tag{5.128}$$

The general solution is:

$$u(\rho) = Ae^{-\rho} + Be^{\rho} \tag{5.129}$$

since $e^\rho \to \infty$ as $\rho \to \infty$, thus $B = 0$, and

$$u(\rho) = Ae^{-\rho} \tag{5.130}$$

However, in Equation (5.127), for $\rho \to 0$, the angular momentum term dominates, and approximately the equation reduces to the following form:

$$\frac{d^2u}{d\rho^2} + \left[-\frac{l(l+1)}{\rho^2} \right] u = 0 \tag{5.131}$$

The general solution is:

$$u(\rho) = C\rho^{l+1} + D\rho^{-l} \tag{5.132}$$

Since $\rho^{-l} \to \infty$ as $\rho \to 0$, thus $D = 0$, and the solution reduces to:

$$u(\rho) \sim C\rho^{l+1} \tag{5.133}$$

5.3.1.2 Power Series Solution

The previous function does not converge for all values of ρ. Thus, we need to express the function by introducing a new function $y(\rho)$ in terms of power series.

Let

$$u(\rho) = \rho^{l+1} e^{-\rho} y(\rho) \tag{5.134}$$

The function $y(\rho)$ is

$$y(\rho) = \sum_{j=0}^{\infty} b_j \rho^j \tag{5.135}$$

Let us now find the coefficients b_j's by substituting these functions in the radial equation. By substituting Equation (5.134) in the radial equation, we get:

$$\rho \frac{d^2y}{d\rho^2} + 2(l+1-\rho)\frac{dy}{d\rho} + [\rho_0 - 2(l+1)]y = 0 \tag{5.136}$$

By differentiating Equation (5.135), we get:

$$\frac{dy}{d\rho} = \sum_{j=0}^{\infty} j b_j \rho^{j-1} = \sum_{j=0}^{\infty} (j+1) b_{j+1} \rho^j \tag{5.137}$$

and

$$\frac{d^2y}{d\rho^2} = \sum_{j=0}^{\infty} j(j+1) b_{j+1} \rho^{j-1} \tag{5.138}$$

Now, by substituting the previous equations into Equation (5.136), we can obtain the following equation:

$$\sum_{j=0}^{\infty} j(j+1) b_{j+1} \rho^j + 2(l+1) \sum_{j=0}^{\infty} (j+1) b_{j+1} \rho^j - 2 \sum_{j=0}^{\infty} j b_j \rho^j + [\rho_0 - 2(l+1)] \sum_{j=0}^{\infty} b_j \rho^j = 0 \tag{5.139}$$

By equating the coefficients of like powers, the following relationship is obtained:

$$b_{j+1} = \left[\frac{2(j+l+1) - \rho_0}{(j+1)(j+2l+2)}\right] b_j \tag{5.140}$$

The recursion formula determines the coefficient and hence the function $y(\rho)$. Let us see what the form of the previous coefficients is. Just for simplicity, let us get rid of "l" and ρ_0 from the recursion formula. Thus, these coefficients reduce to:

$$b_{j+1} = \left[\frac{2(j)}{(j+1)(j)}\right] b_j = \frac{2}{j+1} b_j \tag{5.141}$$

The equation gives a relationship between the first and the next term: $b_1 = 2b_0, b_2 = b_1$, $b_3 = \frac{2}{3}b_2 \dots$. Thus, b_j can be written in terms of b_0 as:

$$b_j = \frac{2^j}{j!} b_0 \rightarrow y(\rho) = b_0 \sum_{j=0}^{\infty} \frac{2^j}{j!} \rho^j = e^{2\rho} b_0 \tag{5.142}$$

and,

$$u(\rho) = b_0 \rho^{l+1} e^{\rho} \tag{5.143}$$

But now we are back to the asymptotic behavior, which we do not want, that is, $u \rightarrow \infty$ as $\rho \rightarrow \infty$. Thus, we need to do something that can prevent such a situation. The only way possible is to make the coefficient of recursion series to terminate (become zero) at some value of "j," that is, for $j = j_{max}$,

$$b_{j_{max}+1} = \left[\frac{2(j_{max} + l + 1) - \rho_0}{(j_{max} + 1)(j_{max} + 2l + 2)}\right] b_{j_{max}} = 0 \tag{5.144}$$

This is possible only when,

$$2(j_{max} + l + 1) - \rho_0 = 0 \tag{5.145}$$

Let,

$$j_{max} + l + 1 = n \tag{5.146}$$

where n is some positive integer, $n = 1, 2, 3 \dots$, and

$$\rho_0 = 2n \tag{5.147}$$

We know that, $k^2 = \frac{-2mE}{\hbar^2}$, $k = \frac{me^2}{2\pi\hbar^2\rho_0}$ (see the beginning of Section 5.3). Thus,

$$E = -\frac{\hbar^2 k^2}{2m} = -\frac{me^4}{8\pi^2 \varepsilon_0^2 \hbar^2 \rho_0^2} \tag{5.148}$$

Therefore, the allowed energy levels of the hydrogen atom are:

$$E_n = -\left[\frac{m}{2\hbar^2}\left(\frac{e^2}{4\pi\varepsilon_0}\right)^2\right]\frac{1}{n^2} \tag{5.149}$$

The wave function for the hydrogen atom is:

$$\Psi(r, \theta, \phi) = R(r)Y(\theta, \phi) \tag{5.150}$$

$$u(\rho) = \rho^{l+1}e^{-\rho} y(\rho) \tag{5.151}$$

$$u(r) = rR(r) \tag{5.152}$$

Thus,

$$R(r) = \frac{1}{r}\rho^{l+1}e^{-\rho} y(\rho) \tag{5.153}$$

$y(\rho)$ is a polynomial of degree $j_{max} = n - l - 1$ in ρ, here l, n and m are called quantum numbers. Using these numbers as subscripts in radial function, we can rewrite it as:

$$R_{nl}(r) = \frac{1}{r}\rho^{l+1}e^{-\rho} y(\rho) \tag{5.154}$$

For the ground state, $n = 1$ and $l = 0$, the energy of the hydrogen atom is:

$$E_1 = -\left[\frac{m}{2\hbar^2}\left(\frac{e^2}{4\pi\varepsilon_0}\right)^2\right] = -13.6\,\text{eV} \tag{5.155}$$

The radial wave function is:

$$R_{10}(r) = \frac{1}{r}\rho e^{-\rho} y(\rho) \tag{5.156}$$

where, $j_{max} = 1 - 0 - 1 = 0$, and $y(\rho) = \sum_{j=0}^{j_{max}} b_j\rho^j = b_0$. Thus,

$$R_{10}(r) = \frac{1}{r}\rho e^{-\rho} b_0 \tag{5.157}$$

It turns out that since, $k^2 = \frac{-2mE}{\hbar^2}$, $k = (\frac{me^2}{4\pi\varepsilon_0\hbar^2})\frac{1}{n}$, for ground state $k = (\frac{me^2}{4\pi\varepsilon_0\hbar^2}) = \frac{1}{a}$, where a is called *Bohr radius*. As $\rho = kr = \frac{r}{a}$, the radial function becomes,

$$R_{10}(r) = \frac{1}{a}e^{-(r/a)} b_0 \tag{5.158}$$

All we need to do is to find the coefficient b_0 using normalization condition for the radial wave function.

$$\int_0^\infty |R_{10}|^2 r^2\, dr = \frac{|b_0|^2}{a^2}\int_0^\infty e^{-\frac{2r}{a}}r^2\, dr = 1 \tag{5.159}$$

By integrating the previous function, the coefficient, $b_0 = \frac{2}{\sqrt{a}}$.
Also, for the ground state, $l = 0$, $m = 0$, and the angular function is:

$$Y_{lm} = Y_{00} = \frac{1}{\sqrt{4\pi}} \tag{5.160}$$

Therefore, the ground state wave function is:

$$\Psi(r, \theta, \phi) = \Psi_{nlm} = R_{nl}(r)Y_{lm}(\theta, \phi) \rightarrow \Psi_{100} = R_{10}(r)Y_{00}(\theta, \phi) = \frac{1}{\sqrt{\pi a^3}}e^{-r/a} \tag{5.161}$$

TABLE 5.3

The Mathematical Form of the Radial Functions for Different Energy States

$n = 1$ $R_{10} = 2a^{-3/2}e^{-r/a}$

$n = 2$ $R_{20} = \frac{1}{\sqrt{2}}a^{-\frac{3}{2}}\left(1 - \frac{1}{2}\frac{r}{a}\right)e^{-r/2a}$

                $R_{21} = \frac{1}{\sqrt{24}}a^{-\frac{3}{2}}\left(\frac{r}{a}\right)e^{-r/2a}$

$n = 3$ $R_{30} = \frac{2}{\sqrt{27}}a^{-\frac{3}{2}}\left(1 - \frac{2}{3}\frac{r}{a} + \frac{2}{27}\left(\frac{r}{a}\right)^2\right)e^{-r/3a}$

                $R_{31} = \frac{8}{27\sqrt{6}}a^{-\frac{3}{2}}\left(1 - \frac{1}{6}\frac{r}{a}\right)\left(\frac{r}{a}\right)e^{-r/3a}$

                $R_{32} = \frac{4}{81\sqrt{30}}a^{-\frac{3}{2}}\left(\frac{r}{a}\right)^2 e^{-r/3a}$

The radial functions for the other higher energy states can be derived in the similar way but make sure that you find the expansion coefficient using the normalization condition, as previously shown. Table 5.3 gives the radial function $R_{nl}(r)$ for higher energy states.

5.3.1.3 The Laguerre Polynomials and the Associated Laguerre Polynomials

The polynomial $y(\rho)$ described in Equation (5.129) and its coefficients described in the recursion formula in Equation (5.132) can be described in a simplified form using Laguerre polynomials as:

$$y(\rho) = L^{2l+1}_{n-l-1}(2\rho) \tag{5.162}$$

where

$$L^{b}_{a-b}(x) = (-1)^b \left(\frac{d}{dx}\right)^a L_a(x) \tag{5.163}$$

is an associated Laguerre polynomial, and

$$L_a(x) = e^x \left(\frac{d}{dx}\right)^a (e^{-x}x^a) \tag{5.164}$$

is the qth Laguerre polynomial. Table 5.4 describes a few of these polynomials.

The complete normalized wave function is:

$$\Psi_{nlm} = \sqrt{\left(\frac{2}{na}\right)^3 \frac{(n-l-1)!}{2n[(N+l)!]^3}} \, e^{-r/na} \left(\frac{2r}{na}\right)^l \left[L^{2l+1}_{n-l-1}\left(\frac{2r}{na}\right)\right] Y^m_l(\theta, \varphi) \tag{5.165}$$

Here, we have not discussed the derivation of normalization of the previous function as it is not very important and is mainly a mathematical procedure.

TABLE 5.4

The Mathematical Form of the Laguerre Polynomials and Associated Laguerre Polynomials

Laguerre Polynomials $L_a(x)$	Associated Laguerre Polynomials $L^b_{a-b}(x)$
$L_0(x) = 1$	$L^0_0 = 1,\ L^1_0 = 1,\ L^2_0 = 2$
$L_1(x) = 1 - x$	$L^0_1 = 1 - x,\ L^1_1 = -2x + 4,\ L^2_1 = -6x + 18$
$L_2(x) = x^2 - 4x + 2$	$L^0_2 = x^2 - 4x + 2,\ L^1_2 = 3x^2 - 18x + 18,\ L^2_2 = 12x^2 - 96x + 144$

In general, the position probability for the electron situated at point (r, θ, ϕ) in the volume element $d^3r = r^2\, dr\, d\Omega$ is given as:

$$dP_{n,l,m} = |\Psi_{nlm}|^2 r^2\, dr\, d\Omega = |R_{n,l}|^2 r^2\, dr \; X \; |Y_l^m(\theta, \phi)|^2 \sin\theta\, d\theta\, d\phi \qquad (5.166)$$

The probability of finding the electron between r and $r + dr$ inside the solid angle under consideration is simply proportional to $|R_{n,l}|^2\, r^2\, dr$. The corresponding probability density is therefore, $|R_{n,l}|^2\, r^2$. The probability density (Equation 5.166) for different energy states of the hydrogen are depicted in Figure 5.9.

EXAMPLE 5.4

Find $\langle r \rangle$ and $\langle r^2 \rangle$ for an electron in ground state of a hydrogen atom.
 The ground state wave function of the hydrogen atom is (Figure 5.9),

$$\Psi_{100} = \frac{1}{\sqrt{\pi a^3}} e^{-r/a} \qquad (5.167)$$

$$\langle r \rangle = \int \Psi_{100} r \Psi_{100}^* r^2 \sin\theta\, d\theta\, d\phi \qquad (5.168)$$

$$= \frac{1}{\pi a^3} \int_0^\infty r^3 e^{-2r/a}\, dr \int_0^{2\pi} d\phi \int_0^\pi \sin\theta\, d\theta$$

$$= \frac{12}{a^3} \left(\frac{a}{2}\right)^4 = \frac{3}{2} a \qquad (5.169)$$

and,

$$\langle r^2 \rangle = \frac{1}{\pi a^3} \int_0^\infty r^4 e^{-2r/a}\, dr \int_0^{2\pi} d\phi \int_0^\pi \sin\theta\, d\theta = 3a^2 \qquad (5.170)$$

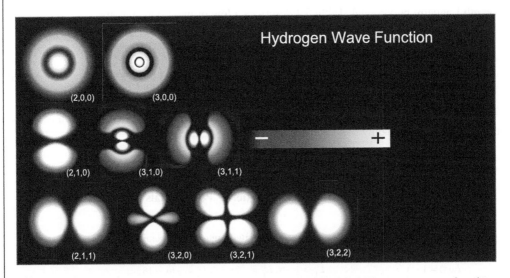

FIGURE 5.9 This image depicts the probability density distribution of the hydrogen atom wavefunctions for different energy states (n, l, m). These images are produced by Haocheng Yin.

5.3.1.4 Degeneracy of Hydrogen Atom

The energy levels of the hydrogen atom are only dependent on the principal quantum number n and not on other quantum numbers l and m. Thus, by knowing the principle quantum number, the energy levels of the hydrogen atom can be determined. For any value of principle quantum number n, the possible values of l are, $l = 0, 1, 2, 3...n - 1$. For each l, there are $(2l + 1)$ possible values of m. This means that many quantum states have the same energy, and therefore, the system is degenerate. For ground state, $n = 1$, $l = 0$, and $m = 0$. For $n = 2$, energy state, l has two values, $l = 0, 1$ and for each l, there are corresponding m values. For $l = 0$, $m = 0$, and for $l = 1$, there are three values of m, as $m = -1, 0, +1$. Therefore, the total degeneracy of the energy level E_n is:

$$D(n) = \sum_{l=0}^{n-1} (2l + 1) = 2\frac{(n - 1)n}{2} + n = n^2 \tag{5.171}$$

This property of energy dependence mainly on the principle quantum number is characteristic of the hydrogen atom only and is due to the nature of the coulombic potential. In other atoms, energy is dependent not only on "n" but also on the "l" quantum number, since their potential function deviates from the coulombic potential due to screening effect of other electrons.

Historically, the principle quantum number "n" is called a shell and is denoted as s. Each shell contains n subshells corresponding to the quantum number "l." Each subshell has $(2l + 1)$ states associated with the quantum number "m." These energy states of the hydrogen atom are depicted in Figure 5.10.

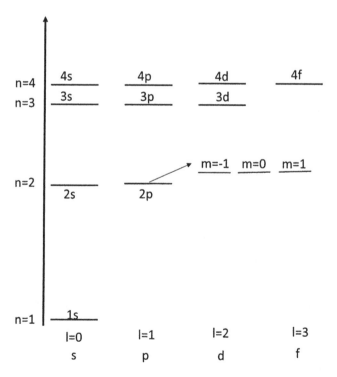

FIGURE 5.10 Energy states distribution of hydrogen atom, shell and subshells are depicted.

5.3.2 HYDROGEN ATOM SPECTRUM

The emission or absorption spectrum of hydrogen gas and many other gases (He, Ne, Xe, etc.) was a subject of intense research for more than two decades at the beginning of the twentieth century. In fact, observation of such a spectrum in the scientific laboratories was the observation of the earliest "quantum effect." Interestingly, no one had a clue about what caused such a spectrum, and scientists were puzzled over its existence. The explanation for its origin came later and laid the foundation of the preliminary version of the quantum mechanics. The learning of quantum mechanics will be incomplete without knowing the story behind the spectrum of hydrogen gas, so let us begin here with a short story on the origin of hydrogen atom spectrum....

It was George Claude in 1910, a French chemist and inventor, who invented the first electrical discharge tube, known today simply as neon light. An electrical discharge tube is a sealed glass tube with a metal electrode at each end, filled with gases such as neon, helium or xenon at low pressure. When voltage is applied between the electrodes, the tube glows with a color that depends on the type of gas contained in the tube. Hydrogen gas in the tube gives a pink color, neon gas gives an orange-to-red color, and helium gives white, gray or blue colors. What causes such a glow and different colors from different gases? The scientists at that time were struggling to explain this phenomenon of the electric discharge tube. It took a decade or so to fully understand it. Now, using quantum mechanics principles, this phenomenon can be easily explained, and it led to the understanding of the hydrogen atom spectrum and the spectrum of many other gases. We will apply the understanding of the structure of the hydrogen atom that we have learned so far for explaining this phenomenon.

As discussed in the previous sections, the energy levels of the hydrogen atom are discrete. According to quantum mechanics, these energy levels are given by the following equation:

$$E_n = -\left[\frac{m}{2\hbar^2}\left(\frac{e^2}{4\pi\varepsilon_0}\right)^2\right]\frac{1}{n^2} \tag{5.172}$$

When an electron jumps from a higher-energy level to a lower-energy level, it emits a photon that carries energy that is equal to the energy of the difference between the two levels and is given as:

$$E_{photon} = -13.6\,\text{eV}\left[\frac{1}{n_f^2} - \frac{1}{n_i^2}\right] \tag{5.173}$$

Going back to the discharge tube, when voltage is applied between the electrodes of the discharge tube, the H_2 molecules of the gas dissociate into hydrogen atoms, and such atoms then collide with each other and with the walls of the tube, which causes electrons in the hydrogen atoms to jump from a higher-energy level to a lower-energy level, emitting radiation or photons that carry energy equivalent to the energy difference between the levels. The glow and color of the radiation observed in the tube is due to the most intense radiation emitted by these electrons. When the light coming through the discharge tube is made to pass through a prism, the spectrum of the light is observed. For hydrogen gas, there are four narrow discrete bands or lines, as shown in the figure below. Each line corresponds to the radiation emitted by electrons from a higher level to a lower level, as shown in the Figure 5.11. This line spectrum is very distinct from the spectrum of continuous bands of colors obtained from the light of a hot filament. This observation of the line spectrum instead of the continuous band spectrum of the light coming from discharge tube puzzled scientists at that time. The discreteness in the spectrum of the narrow bands mainly arises due to discreteness in the atomic structure of the atoms. The atomic structure of an atom is its signature characteristic, which is revealed in the line spectrum. Every atom is completely distinct from the other atoms due to its atomic structure. Thus, by looking at the line spectrum of the light, the information about the source can be revealed. This is precisely how the source of light coming from the stars was identified, and the first lines of hydrogen spectra were found not in laboratory but in the spectra of light from stars. In 1881, Sir William Huggins identified 10 lines of hydrogen emission spectra by looking at the first photographs of stellar spectra. These lines varied in wavelength from the red visible light to the near ultraviolet light. In 1885, working from astronomical measurements, Johann Jakob Balmer found that he could account for the positions of all known lines by applying a simple

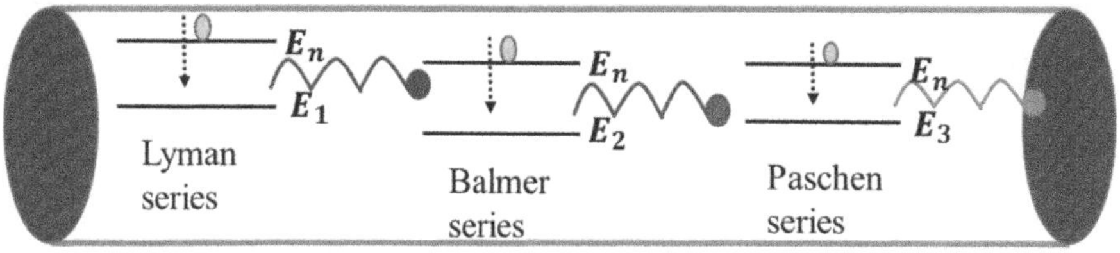

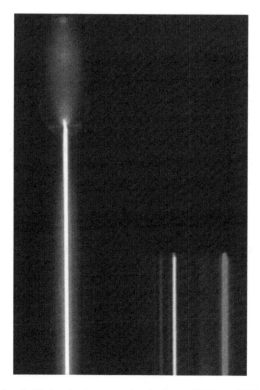

FIGURE 5.11 Electrical discharge tube or simply "neon light" showing several electronic transitions in hydrogen atoms taking place, causing observation of line spectrum.

empirical formula. The entire set of lines has come to be known as Balmer series. Another group of series in the far ultraviolet region are called Lyman series and Paschen series. The Balmer series arises due to transitions taking place from any higher level "n" to a lower level 2, as shown in the Figure 5.11. All lines in the series share the same lower state. Similarly, the Lyman and Paschen series also share the same lower states. These lines can be further split into closer lines that result mainly from relativistic and magnetic interactions related to orbital and spin angular momentum of the electrons of the atom. This splitting is called fine structure of the spectrum. There is so much that can be revealed about the hydrogen atom by studying it atomic spectrum. New findings about the hydrogen spectrum will continue as news tools are becoming available for obtaining spectrum. For further details about the spectrum of hydrogen, a reference is mentioned in references section of the book (Figures 5.11 and 5.12).

FIGURE 5.12 On the left hand side is a hydrogen spectral tube excited by a 5000-volt transformer. The three prominent lines observed from the spectra of the radiation obtained from the tube are shown at the right of the image through a 600 lines/mm diffraction grating. These are Balmer lines in the visible range of the spectrum. (This image is taken from HyperPhysics website http://hyperphysics.phy-astr.gsu.edu/hbase/hyde.html, Credit: Dr. Rod Nave.)

Conceptual Question 7: *Explain the difference between line spectrum and continuous band spectrum.*

PROBLEMS

5.11 Determine the radial wave functions R_{20} and R_{21}. Using these radial functions, find the normalized wave functions Ψ_{200} and Ψ_{210}.

5.12 Obtain the mathematical expression for the probability densities of the electron in 1s, 2s and 2p shells. Plot their corresponding radial probability densities.

5.13 Find the most probable value of radius "r" in the 1 and 2s shells. Find the expectation value of the potential energy of electron in these shells.

5.14 Prove that the spherical harmonics are orthogonal functions.

5.15 Consider that the electron of a hydrogen atom is in the ground state. Calculate the expectation value of its position from the nucleus. Does this expectation value changes with time?

5.16 Find the number of shells and subshells associated with principle quantum number $n = 5$ for the hydrogen atom.

5.17 What was the earliest quantum effect related to atomic spectra observed in the laboratories? Describe in your own words. What is the difference between Lyman, Balmer and Paschen series?

5.4 THE ANGULAR MOMENTUM

In classical mechanics, angular momentum of a system or body is related to its rotation in space. The angular momentum is a constant of motion when the net external torque acting on the system is zero. Thus, for a closed system, the angular momentum of a body is a conserved quantity due to isotropy of space. This isotropy means that the angular momentum of a body in a closed system does not vary when it is rotated as a whole in any manner in space. Generally, energy, momentum and angular momentum of a closed system are conserved quantities. Classically, the angular momentum of a particle of mass "m" is defined as:

$$\vec{L} = \vec{r} \times \vec{p} \tag{5.174}$$

Here, $\vec{r}$ is the radius vector that describes the location of the particle in space, and $\vec{p}$ is the linear momentum.

$$\vec{L} = \begin{vmatrix} \hat{i} & \hat{j} & \hat{k} \\ x & y & z \\ p_x & p_y & p_z \end{vmatrix}$$

Thus,

$$L_x = yp_z - zp_y, \quad L_y = zp_x - xp_z, \quad L_z = xp_y - yp_x \tag{5.175}$$

The commutation relationships for above variables can be described as:

$$[L_x, L_y] = L_z, \quad [L_y, L_z] = L_x, \quad [L_z, L_x] = L_y \tag{5.176}$$

In classical mechanics, the equation of motion of a dynamical variable (angular momentum) is:

$$\frac{d\vec{L}}{dt} = [H, \vec{L}] \tag{5.177}$$

If,

$$[H, \vec{L}] = 0 \tag{5.178}$$

Then,

$$\frac{d\vec{L}}{dt} = 0 \tag{5.179}$$

$$\vec{L} = \text{const.} \tag{5.180}$$

Thus, angular momentum L is a constant of motion and is a conserved quantity. In classical mechanics, any dynamical variable of a system that commutes with the Hamiltonian is a constant of motion.

EXAMPLE 5.5

Consider the Hamiltonian of a spherically symmetric potential $V(r)$:

$$H = \frac{p^2}{2m} + V(r) \tag{5.181}$$

The angular momentum of a particle that is rotating under the influence of the previous potential will be conserved. Let us see why:

$$[H, L] = \frac{1}{2m}[H, L] = \frac{1}{2m}\left[[p^2, L_x] + [p^2, L_y] + [p^2, L_z]\right] + [V(r), L] \tag{5.182}$$

$$\left[p_x^2, L_x\right] = p_x[p_x, L_x] + [p_x, L_x]p_x \tag{5.183}$$

$$[p_x, L_x] = [p_x, yp_z - zp_y] = y[p_x, p_z] + [p_x, y]p_z - [p_x, z]p_y - z[p_x, p_y] = 0 \tag{5.184}$$

$$[p_x, L_y] = [p_x, zp_x - xp_z] = z[p_x, p_x] + [p_x, z]p_x - [p_x, x]p_z - x[p_x, p_z] = p_z \tag{5.185}$$

$$[p_x, L_z] = [p_x, xp_y - yp_x] = [p_x, x]p_y + x[p_x, p_y] - [p_x, y]p_x - y[p_x, p_x] = -p_y \tag{5.186}$$

$$[p_y, L_x] = -p_z, \quad [p_z, L_x] = p_y \tag{5.187}$$

$$[p^2, L_x] = \left[p_x^2, L_x\right] + \left[p_y^2, L_x\right] + \left[p_z^2, L_x\right] \tag{5.188}$$

$$\left[p_x^2, L_x\right] = 0 \tag{5.189}$$

$$\left[p_y^2, L_x\right] = -p_y p_z - p_z p_y \tag{5.190}$$

$$\left[p_z^2, L_x\right] = p_z p_y + p_y p_z \tag{5.191}$$

Thus,

$$[p^2, L_x] = [p_x^2, L_x] + [p_y^2, L_x] + [p_z^2, L_x] = -p_y p_z - p_z p_y + p_z p_y + p_y p_z = 0 \qquad (5.192)$$

Similarly,

$$[p^2, L_y] = [p_x^2, L_y] + [p_y^2, L_y] + [p_z^2, L_y] = 0 \qquad (5.193)$$

$$[p^2, L_z] = [p_x^2, L_z] + [p_y^2, L_z] + [p_z^2, L_z] = 0 \qquad (5.194)$$

Hence,

$$[H, L] = \frac{1}{2m}[H, L] = \frac{1}{2m}\left[[p^2, L_x] + [p^2, L_y] + [p^2, L_z]\right] + [V(r), L] = 0 \qquad (5.195)$$

Because,

$$[V(r), L] = r \times \nabla V = r \times \hat{r}\frac{\partial V}{\partial r} = 0 \qquad (5.196)$$

According to the equation of motion for the angular momentum,

$$\frac{d\vec{L}}{dt} = [H, \vec{L}] = 0 \qquad (5.197)$$

Therefore, L is a constant of motion and is a conserved quantity.

In Section 5.2, the angular momentum quantum numbers l and m were discussed. What is the meaning of angular momentum in quantum mechanics? We will try to answer these questions in this section.

For a hydrogen atom, an electron orbiting the nucleus can be considered as a particle undergoing rotation about the axis passing through the nucleus as center. Such a particle has angular momentum as L and whose components are L_x, L_y, and L_z. These angular momentums in operator form can be described as:

$$\hat{L}_x = \frac{\hbar}{i}\left(-\sin\phi\,\frac{\partial}{\partial\theta} - \cos\phi\cot\theta\,\frac{\partial}{\partial\phi}\right) \qquad (5.198)$$

$$\hat{L}_y = \frac{\hbar}{i}\left(\cos\phi\,\frac{\partial}{\partial\theta} - \sin\phi\cot\theta\,\frac{\partial}{\partial\phi}\right) \qquad (5.199)$$

$$\hat{L}_z = \frac{\hbar}{i}\frac{\partial}{\partial\phi} \qquad (5.200)$$

Raising and lowering operators are:

$$\hat{L}_\pm = \hat{L}_x \pm i\,\hat{L}_y \qquad (5.201)$$

$$\hat{L}_\pm = \pm\hbar e^{\pm i\phi}\left(\frac{\partial}{\partial\theta} \pm i\cot\theta\,\frac{\partial}{\partial\phi}\right) \qquad (5.202)$$

$$\hat{L}^2 = \hat{L}_x^2 + \hat{L}_y^2 + \hat{L}_z^2 = -\hbar^2\left[\frac{1}{\sin\theta}\frac{\partial}{\partial\theta}\left(\sin\theta\frac{\partial}{\partial\theta}\right) + \frac{1}{\sin^2\theta}\left(\frac{\partial^2}{\partial\phi^2}\right)\right] \qquad (5.203)$$

5.4.1 What Are the Eigenfunctions and Eigenvalues of These Operators?

As was discussed in Section 5.2.3, the solution of the angular equation for $\hat{L}^2$ gives the eigenvalues and eigenfunctions that are spherical harmonics ($Y_l^m(\theta, \phi)$).

$$\hat{L}^2 f_l^m = \hbar^2 l(l+1) f_l^m \tag{5.204}$$

It is also an eigenfunction of the operator

$$\hat{L}_z f_l^m = \hbar m \, f_l^m \tag{5.205}$$

Here, function f_l^m is same as spherical harmonics ($Y_l^m(\theta, \phi)$). Thus, spherical harmonics are simultaneous eigenfunctions of $\hat{L}_z$ and $\hat{L}^2$.

Revisiting the Hamiltonian operator described in Section 5.2,

$$\hat{H} = -\frac{\hbar^2}{2m} \left\{ \frac{1}{r^2} \frac{\partial}{\partial r} \left(r^2 \frac{\partial}{\partial r} \right) \right\} + \frac{\hat{L}^2}{2mr^2} + V(r) \tag{5.206}$$

Here, it can be seen that angular momentum operator L^2, and H will have the same energy eigenfunctions (spherical harmonics) and thus will commute:

$$[\hat{H}, \hat{L}^2] = 0 \tag{5.207}$$

Similarly,

$$[\hat{H}, \hat{L}] = 0 \tag{5.208}$$

Therefore, $\hat{L}^2$ and $\hat{L}$ are constants of motion.

Let us discuss the commutation relationships of $\hat{L}_x$, $\hat{L}_y$ and $\hat{L}_z$. In quantum mechanics, these commutation relationships are:

$$[\hat{L}_x, \hat{L}_y] = i\hbar \hat{L}_z, [\hat{L}_y, \hat{L}_z] = i\hbar \hat{L}_x, [\hat{L}_z, \hat{L}_x] = i\hbar \hat{L}_y \tag{5.209}$$

Thus, $\hat{L}_x$, $\hat{L}_y$ and $\hat{L}_z$, are incompatible observables (noncommutative) and do not have simultaneous eigenfunctions. Only $\hat{H}$, $\hat{L}^2$ and $\hat{L}_z$ have simultaneous eigenfunctions.

The Heisenberg uncertainty relationship for these observables can be described as:

$$\sigma_x \, \sigma_y \geq \frac{\hbar}{2} |\langle \hat{L}_z \rangle| \tag{5.210}$$

where

$$\sigma_x^2 = \langle (\hat{L}_x - \langle \hat{L}_x \rangle)^2 \rangle, \quad \sigma_y^2 = \langle (\hat{L}_y - \langle \hat{L}_y \rangle)^2 \rangle \tag{5.211}$$

which means that observables $\hat{L}_x$, $\hat{L}_y$ and $\hat{L}_z$ cannot simultaneously have definite values. If $\hat{L}_z$ has a well-defined value, then $\hat{L}_x$ and $\hat{L}_y$ do not. If $\hat{L}_x$ has a well-defined value, then $\hat{L}_y$ and $\hat{L}_z$ do not. Therefore, a quantum particle rotating in space cannot have a determinate angular momentum.

5.4.2 Is Angular Momentum Conserved in a Quantum Mechanical System?

According to the principle of uncertainty, all three components of the angular momentum cannot be simultaneously determined. It is not that we do not have best devices that cannot measure the angular momentum components simultaneously. It is just that for a quantum mechanical system, if we know the angular momentum of one of the components, then we are completely uncertain about the other two. In simple words, the other two components do not even exist, and without the other two components, we cannot know the angular momentum of the system. Thus, for a quantum mechanical system, since we are uncertain about the angular momentum, we cannot be certain about the conservation of angular momentum as we are in classical mechanics. In the following example, the closest meaning of conservation of angular momentum in quantum mechanics is discussed.

EXAMPLE 5.6

Consider the following Hamiltonian, $\hat{H} = \frac{\hat{p}^2}{2m} + V(r)$, show that,

$$\frac{d\langle \hat{L} \rangle}{dt} = 0 \tag{5.212}$$

Using expectation value equation for the x-component of the angular momentum,

$$\frac{d\langle \hat{L}_x \rangle}{dt} = \frac{i}{\hbar} \langle [\hat{H}, \hat{L}_x] \rangle \tag{5.213}$$

$$[\hat{H}, \hat{L}_x] = \frac{1}{2m} \left[[\hat{p}^2, \hat{L}_x] \right] + [\hat{V}, \hat{L}_x] \tag{5.214}$$

$$[\hat{H}, \hat{L}_x] = \frac{1}{2m} \left[[\hat{p}_x^2, \hat{L}_x] + [\hat{p}_y^2, \hat{L}_x] + [\hat{p}_z^2, \hat{L}_x] \right] + [\hat{V}, \hat{L}_x] \tag{5.215}$$

$$[\hat{p}^2, \hat{L}_x] = [\hat{p}_x^2, \hat{L}_x] + [\hat{p}_y^2, \hat{L}_x] + [\hat{p}_z^2, \hat{L}_x] = 0 \tag{5.216}$$

and

$$\begin{aligned} [\hat{V}, \hat{L}_x] &= [\hat{V}, y\hat{p}_z - z\hat{p}_y] = [\hat{V}, y\hat{p}_z] - [\hat{V}, z\hat{p}_y] \\ &= y[\hat{V}, \hat{p}_z] + [\hat{V}, y]\hat{p}_z - z[\hat{V}, \hat{p}_y] - [\hat{V}, z]\hat{p}_y \end{aligned} \tag{5.217}$$

$$[\hat{V}, \hat{p}_z] = -\frac{\hbar}{i}\frac{\partial V}{\partial z}, \quad [\hat{V}, \hat{p}_y] = -\frac{\hbar}{i}\frac{\partial V}{\partial y} \tag{5.218}$$

$$[\hat{V}, \hat{L}_x] = y[\hat{V}, \hat{p}_z] - z[\hat{V}, \hat{p}_y] = -\frac{\hbar}{i}\left[y\frac{\partial V}{\partial z} - z\frac{\partial V}{\partial y} \right] = -\frac{\hbar}{i}[r \times \nabla V]_x \tag{5.219}$$

Similarly, other two components can be obtained in the same way.

Thus, for $V = V(r)$ which is a spherically symmetric potential,

$$\frac{d\langle \hat{L} \rangle}{dt} = \langle [r \times \nabla V] \rangle = 0 \tag{5.220}$$

because $\nabla V = \frac{\partial V}{\partial r}\hat{r}$ and $\vec{r} \times \hat{r} = 0$.

This implies that the expectation value of angular momentum remains constant in time. Average angular momentum is a conserved quantity. Therefore, quantum mechanically, angular momentum is related to $\hat{L}^2$ and $\hat{L}_z$ or any other one of the components of angular momentum. In physics, any physical quantity or dynamical variable that cannot be determined or measured experimentally has no meaning. Since we cannot measure all the three components of angular momentum simultaneously, angular momentum itself has no meaning, but $\hat{L}^2$ and $\hat{L}_z$ remain meaningful quantities. Thus, we describe the atomic shells in terms of the square of the total angular quantum number "l," *which is a positive integer or half integer*, and the magnetic quantum number "m," which takes any value between $-l, l+1,..., l-1, l$. For a given value of l, there are $2l+1$ values of m.

Angular momentum is a very well-understood concept in classical mechanics. Rotations of rigid bodies and celestial objects like the Earth are based on this important understanding, but here the meaning of angular momentum in quantum mechanics has surprised us in similar ways as position and momentum.

Conceptual Question 8: Explain the uncertainty in the measurement of angular momentum of a quantum system.

PROBLEMS

5.18 The raising and lowering operators change the value of "m" by one. Find $\hat{L}_\pm Y_1^1(\theta, \phi)$. Does this operator play the same role of a raising operator like that of a harmonic oscillator?

5.19 Derive expressions for $[\hat{V}, \hat{L}_y]$, and $[\hat{V}, \hat{L}_z]$.

5.20 What is the difference between the classical conservation of momentum and Equation (5.220)? Is the angular momentum of a quantum particle a conserved quantity?

5.21 Using commutation relationships shown in Equation (5.175), prove the following commutation relationships, $[\hat{L}_z, x] = i\hbar y$, and $[\hat{L}_z, y] = -i\hbar x$.

5.22 Show that $[\hat{L}^2, \hat{L}] = 0$.

5.5 TUTORIAL

5.5.1 ANGULAR MOMENTUM

Purpose

To understand the difference between classical and quantum angular momentum.

Concepts

- Classically, angular momentum is related to the rotation of a body in space and is constant or conserved when the net external torque acting on the system is zero.
- It is described as a three-dimensional vector. The Hamiltonian of any classical system performing rotation in space and under the influence of radial potential can be described as:

$$H = \frac{p_r^2}{2m} + \frac{L^2}{2mr^2} + V(r)$$

where p_r is the radial angular momentum, and L is the angular momentum vector. All three components of the angular can be determined simultaneously for any system.

- Angular momentum of a quantum system is discrete and is described by quantum number "l," which is a positive integer or half integer, and magnetic quantum number "m," which takes any value between $-l, l+1, \ldots, l-1, l$. For a given value of l, there are $2l+1$ values of m.
- Quantum mechanically, the angular momentum (L), the square of angular momentum (L^2), and one of the angular momentum (L_z) can only be determined simultaneously. Thus, these two quantities have physical meaning. However, if one of the components of angular momentum is known definitely, then other two components become indeterminate.

Listen to the following conversation between two friends, Classico and Quantix, and answer the questions that follow.

Classico: "If somebody asked me to describe the dimensions of my table, I will tell them that the height of my table is $1m$, length $1/2m$ and width $1/4m$. Thus, the size vector (M) of my table can be described as:

$$\vec{M} = 1m\,\hat{i} + 0.5m\,\hat{j} + 0.25m\,\hat{k}.$$

The magnitude of this vector is:

$$\left|\vec{M}\right| = \sqrt{1m^2 + 0.25m^2 + 0.06m^2}$$

But after studying the angular momentum in quantum mechanics class today, this question came to my mind. If I do not know the length and the width of the table, can I know the magnitude of the size vector of the table?"

Quantix: "Well, according to quantum mechanics, yes you can. Even if you do not know the other two components of the size vector, the square of the size vector (M^2) can still be determined. You may never know the exact length and width of your table, but you can still know the magnitude of square of the size vector."

Classico: "It sounds very puzzling to me. This is a violation of the fundamental laws."

Quantix: "According to quantum mechanics, all tables are not made the same way. If you are buying two tables of the same size, they may still be of different dimensions."

Question 1 Do you agree/disagree with Classico or Quantix? Explain using the angular momentum concept.

_____

_____

_____

_____

_____

Question 2 Do you agree with Classico, when he says, "This is a violation of the fundamental laws." Is it true?

_____

_____

_____

_____

_____

Question 3 Why does the angular momentum of a quantum mechanical system not have any physical meaning? Explain in your own words.

_____

_____

_____

_____

_____

Question 4 Explain the differences between the following two equations describing the classical and quantum Hamiltonians,

Classical Hamiltonian	Quantum Hamiltonian
$H = \frac{p_r^2}{2m} + \frac{L^2}{2mr^2} + V(r)$	$\hat{H} = -\frac{\hbar^2}{2m}\left\{\frac{1}{r^2}\frac{\partial}{\partial r}\left(r^2\frac{\partial}{\partial r}\right)\right\} + \frac{\hat{L}^2}{2mr^2} + V(r)$

_____

_____

_____

_____

_____

Question 5   Write the wave function of the previous quantum mechanical Hamiltonian. Using this wave function, find the wave function of an electron in the ground state of the hydrogen atom.

_____

_____

_____

_____

_____

Question 6   Suppose we wish to measure experimentally the angular momentum of the electrons in a given atom. A beam of such atoms is then prepared and made to pass through a set of collimating slits as shown in the following Figure 5.13:

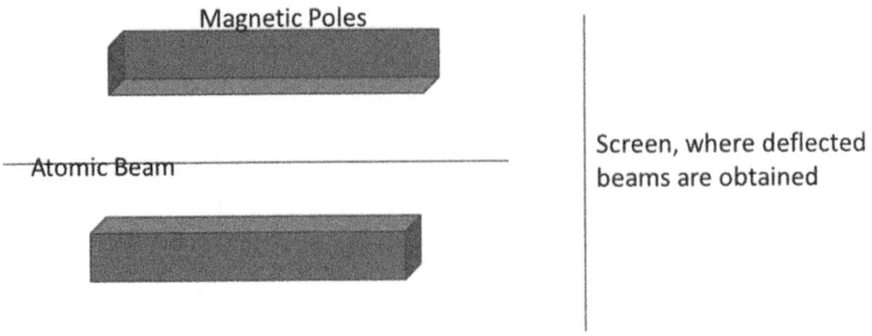

_____

_____

_____

_____

_____

FIGURE 5.13   The illustration of an atomic beam passing through magnetic poles.

The beam enters the region in which there is a magnetic field that is normal to the direction of the motion of atoms. In the homogenous field, the atoms experiences force:

$$F = \nabla(\mu.B)$$

where μ is the magnetic moment of the electrons in the atom, and B is the magnetic field.

$$\mu = \frac{eL}{2mc}$$

Show that each atom experiences a force that is proportional to the z-component of the electronic angular momentum.

The beam is collected some distance from the magnet at a point far enough away so that atoms of different L_z have been separated. Explain how by measuring the deflection of the beam that the angular momentum L_z of the atom can be calculated.

_____

_____

_____

_____

_____

Draw a diagram showing all the deflected beams on the screen.

Classically, what should be observed instead of the deflected beams? Draw a diagram.

6 Perturbation Theory

6.1 TIME-INDEPENDENT PERTURBATION

In the previous chapters, we have shown how to solve the Schrodinger equation for several systems as a particle in an infinite and finite potential well, scattering, harmonic oscillator, and hydrogen atom. For all these systems, we were able to solve Schrodinger's equation exactly. However, there are complicated potentials for which Schrodinger's equation cannot be solved exactly. For such situations, we resort to approximation methods for finding approximate solutions to the Schrodinger equation. In this chapter, we discuss several approximate methods for solving Schrodinger's equation.

Consider a particle in an infinite square well. For such a system, we exactly know the solution of the Schrodinger equation. Now, consider that a perturbation is being applied that changes slightly the form of the infinite potential function. We need to find the solution of Schrodinger's equation for such a perturbed system. The method that is commonly used for such a system is called perturbation theory. Such a method is based on systematically obtaining an approximate solution to the perturbed system by developing solutions based on the exact solutions of the unperturbed case.

To understand such approach, consider the following Hamiltonian:

$$H = H^0 + \lambda H'$$

(6.1)

where $H' =$ the perturbation, $H^0 =$ the Hamiltonian of the unperturbed system, and $\lambda =$ a small number that varies between 0 and 1. The magnitude of this number decides how strong or how weak the perturbation is. The wave function and energy for the perturbed system can now be written as:

$$\psi_n = \psi_n^0 + \lambda \psi_n^1 + \lambda^2 \psi_n^2 + \lambda^3 \psi_n^3 + \cdots$$

(6.2)

$$E_n = E_n^0 + \lambda E_n^1 + \lambda^2 E_n^2 + \lambda^3 E_n^3 + \cdots$$

(6.3)

Here, ψ_n^1 and E_n^1 are the first-order corrections to the wave function of the nth energy state and the nth energy state. Similarly, ψ_n^2 and E_n^2 are the second-order correction. The Schrodinger equation of the system can be written as:

$$H\psi_n = E_n\psi_n$$

(6.4)

Substituting Equations (6.2) and (6.3) in Equation (6.4), we get:

$$\begin{aligned}
\left[H^0 + \lambda \, H' \psi_n^0 + \lambda \psi_n^1 + \lambda^2 \psi_n^2 + \lambda^3 \psi_n^3 + \cdots \right] \\
= \left[E_n^0 + \lambda E_n^1 + \lambda^2 E_n^2 + \lambda^3 E_n^3 + \cdots \right] \\
\times \left[\psi_n^0 + \lambda \psi_n^1 + \lambda^2 \psi_n^2 + \lambda^3 \psi_n^3 + \cdots \right]
\end{aligned}$$

(6.5)

Conceptual Question 1: What is the significance of the parameter λ?

6.1.1 NON-DEGENERATE PERTURBATION

6.1.1.1 First-Order Correction

By collecting like powers of λ in the previous equation, the first-order correction can be written as:

$$H^0 \psi_n^1 + H' \psi_n^0 = E_n^0 \psi_n^1 + E_n^1 \psi_n^0 \tag{6.6}$$

By taking the inner product with $\langle \psi_n^0 |$, the previous equation is:

$$\langle \psi_n^0 | H^0 | \psi_n^1 \rangle + \langle \psi_n^0 | H' | \psi_n^0 \rangle = \langle \psi_n^0 | E_n^0 | \psi_n^1 \rangle + \langle \psi_n^0 | E_n^1 | \psi_n^0 \rangle \tag{6.7}$$

Since H^0 is Hermitian, the previous equation becomes:

$$E_n^0 \langle \psi_n^0 | \psi_n^1 \rangle + \langle \psi_n^0 | H' | \psi_n^0 \rangle = E_n^0 \langle \psi_n^0 | \psi_n^1 \rangle + E_n^1 \langle \psi_n^0 | \psi_n^0 \rangle \tag{6.8}$$

Thus,

$$E_n^1 = \langle \psi_n^0 | H' | \psi_n^0 \rangle \tag{6.9}$$

The previous equation is a first-order perturbation correction to the energy. Adding this value to the unperturbed energy of the nth level will give the energy of the nth level of the perturbed system. Let us now find the first-order correction to the wave function. First, rewrite Equation (6.6) as:

$$(H^0 - E_n^0) \psi_n^1 = -(H' - E_n^1) \psi_n^0 \tag{6.10}$$

ψ_n^1 is a function that can be expressed as a linear superposition of some other function. Therefore, we choose ψ_n^0, which is a known function of the unperturbed system to be that function. Thus,

$$\psi_n^1 = \sum_{m \neq n} c_m^n \psi_m^0 + c_m^m \psi_m^0 \tag{6.11}$$

In the previous equation, the second term is for the case $n = m$.
Equation (6.10), can be written as:

$$c_m^m (E_m^0 - E_m^0) \langle \psi_l^0 | \psi_m^0 \rangle + \sum_{m \neq n} (E_m^0 - E_n^0) c_m^n \langle \psi_l^0 | \psi_m^0 \rangle = -\langle \psi_l^0 | H' | \psi_n^0 \rangle + E_n^1 \langle \psi_l^0 | \psi_n^0 \rangle \tag{6.12}$$

The first term in the previous equation simply vanishes, and the remaining equations is:

$$\sum_{m \neq n} (E_m^0 - E_n^0) c_m^n \langle \psi_l^0 | \psi_m^0 \rangle = -\langle \psi_l^0 | H' | \psi_n^0 \rangle + E_n^1 \langle \psi_l^0 | \psi_n^0 \rangle \tag{6.13}$$

If $n = l$, the left-hand side is zero, but if $l \neq n$, then the previous equation is:

$$c_m^n = \frac{\langle \psi_m^0 | H' | \psi_n^0 \rangle}{(E_n^0 - E_m^0)} \tag{6.14}$$

Therefore,

$$\psi_n = \psi_n^0 + \lambda c_n^n \psi_n^0 + \lambda \sum_{m \neq n} \frac{\langle \psi_m^0 | H' | \psi_n^0 \rangle}{\left(E_n^0 - E_m^0\right)} \psi_m^0 \tag{6.15}$$

$$\langle \psi_n | \psi_n \rangle = 1 = \left[\langle \psi_n^0 | + \lambda c_n^{n*} \langle \psi_n^0 | + \lambda \sum_{m \neq n} \frac{\langle \psi_m^0 | H' | \psi_n^0 \rangle}{\left(E_n^0 - E_m^0\right)} \langle \psi_m^0 | \right]$$

$$\times \left[|\psi_n^0\rangle + \lambda c_n^n |\psi_n^0\rangle + \lambda \sum_{m \neq n} \frac{\langle \psi_m^0 | H' | \psi_n^0 \rangle}{\left(E_n^0 - E_m^0\right)} |\psi_m^0\rangle | \right].$$

Thus, we obtain:

$$1 = 1 + \lambda c_m^m + \lambda c_m^{m*} + \text{terms in } \lambda^2$$

and therefore,

$c_m^m + c_m^{m*} = 0$ which means it must be a purely imaginary number $i\alpha$. Thus,

$$c_m^m = 1 + i\lambda \alpha + \lambda^2 \text{ terms}$$

$$c_m^m = e^{i\lambda \alpha} = 1 + i\lambda \alpha + \lambda^2 \text{ terms}$$

Hence,

$$\psi_n^1 = e^{i\lambda \alpha} \psi_n^0 + \sum_{m \neq n} \frac{\langle \psi_m^0 | H' | \psi_n^0 \rangle}{\left(E_n^0 - E_m^0\right)} \psi_m^0 \tag{6.16}$$

If the energy states of the system are nondegenerate, the previous equation works. However, the numerator of the previous term even for the nondegenerate case is quite tricky and will give problems. Therefore, the first-order correction to the wave function is not a good approximation.

EXAMPLE 6.1

Consider that a perturbation is being applied to the infinite potential well such that it raises the potential by V_0. The perturbed Hamiltonian of the system can be written as:

$$H' = V_0 \tag{6.17}$$

The unperturbed energy of the infinite square well is:

$$E_n = \frac{n^2 \pi^2 \hbar^2}{2\,mL^2} \tag{6.18}$$

The first-order correction to energy is:

$$E_n^1 = \langle \psi_n^0 | H' | \psi_n^0 \rangle = V_0 \tag{6.19}$$

Thus, the corrected energy is:

$$E_n = \frac{n^2 \pi^2 \hbar^2}{2\,mL^2} + V_0 \tag{6.20}$$

6.1.1.2   Second-Order Correction

For some systems, there are no first-order corrections, and we need to adhere to second-order corrections.

We will follow the same steps as we did previously, and write the second-order equation as:

$$\langle \psi_n^0 | H^0 | \psi_n^2 \rangle + \langle \psi_n^0 | H' | \psi_n^1 \rangle = \langle \psi_n^0 | E_n^0 | \psi_n^2 \rangle + \langle \psi_n^0 | E_n^1 | \psi_n^1 \rangle + \langle \psi_n^0 | E_n^2 | \psi_n^0 \rangle \tag{6.21}$$

Since H^0 is Hermitian, the previous equation becomes:

$$E_n^0 \langle \psi_n^0 | \psi_n^2 \rangle + \langle \psi_n^0 | H' | \psi_n^1 \rangle = E_n^0 \langle \psi_n^0 | \psi_n^2 \rangle + E_n^1 \langle \psi_n^0 | \psi_n^1 \rangle + E_n^2 \langle \psi_n^0 | \psi_n^0 \rangle \tag{6.22}$$

$$\langle \psi_n^0 | H' | \psi_n^1 \rangle - E_n^1 \langle \psi_n^0 | \psi_n^1 \rangle = E_n^2 \tag{6.23}$$

By using Equation (6.16) in the previous equation, we get:

$$e^{-i\lambda\alpha} \langle \psi_n^0 | H' | \psi_n^0 \rangle + \sum_{m \neq n} c_m^n \langle \psi_n^0 | H' | \psi_m^0 \rangle - E_n^1 e^{-i\lambda\alpha} \langle \psi_n^0 | \psi_n^0 \rangle$$
$$- E_n^1 \langle \psi_n^0 | \psi_m^0 \rangle \left[\sum_{m \neq n} \frac{\langle \psi_m^0 | H' | \psi_n^0 \rangle}{\left(E_n^0 - E_m^0 \right)} \right] = E_n^2 \tag{6.24}$$

Since,

$$\langle \psi_n^0 | \psi_n^1 \rangle = \sum_{m \neq n} c_m^n \langle \psi_n^0 | \psi_m^1 \rangle = 0$$

$$E_n^1 e^{-i\lambda\alpha} + \sum_{m \neq n} c_m^n \langle \psi_n^0 | H' | \psi_m^0 \rangle - E_n^1 e^{-i\lambda\alpha} = E_n^2 \tag{6.25}$$

Therefore,

$$E_n^2 = \sum_{m \neq n} c_m^n \langle \psi_n^0 | H' | \psi_m^0 \rangle = \sum_{m \neq n} \frac{\langle \psi_m^0 | H' | \psi_n^0 \rangle \langle \psi_n^0 | H' | \psi_m^0 \rangle}{\left(E_n^0 - E_m^0 \right)} \tag{6.26}$$

$$E_n^2 = \sum_{m \neq n} \frac{|\langle \psi_n^0 | H' | \psi_m^0 \rangle|^2}{\left(E_n^0 - E_m^0 \right)} \tag{6.27}$$

This is a *second-order correction* to energy. Here, we will not go further into finding second-order, third-order and fourth-order corrections to the wave function, but rather, we will discuss the case of degenerate energy.

6.1.2   DEGENERATE PERTURBATION

6.1.2.1   What Is a Degenerate State?

As was discussed in the formalism chapter (Chapter 3), degenerate states are two or more states that have the same energy. If perturbation is applied to such a system, the perturbation theory would fail (as the terms for the first-order correction to the wave function and the second-order correction of energy would blow up) to provide the correct energy and wave function for such a system. Therefore, we must find some new method for dealing with such systems.

Consider following degenerate states of a system:

$$H^0 \psi_\alpha^0 = E^0 \psi_\alpha^0, \quad H^0 \psi_\beta^0 = E^0 \psi_\beta^0, \quad \langle \psi_\alpha^0 | \psi_\beta^0 \rangle = 0, \tag{6.28}$$

The linear superposition of the previous degenerate states can be written as:

$$\psi^0 = a \, \psi_\alpha^0 + b \, \psi_\beta^0 \tag{6.29}$$

Let us now consider the Hamiltonian of the system when a perturbation is applied:

$$H = H^0 + \lambda H' \tag{6.30}$$

with,

$$E = E^0 + \lambda E^1 + \lambda^2 E^2 + \cdots \tag{6.31}$$

$$\psi = \psi^0 + \lambda \psi^1 + \lambda^2 \psi^2 + \cdots \tag{6.32}$$

Substituting the previous equations in Equation (6.30) and collecting like powers of λ, we get:

$$H^0 \psi^1 + H' \psi^0 = E^0 \psi^1 + E^1 \psi^0 \tag{6.33}$$

Taking the inner product of the previous equation with ψ_α^0, we get the following:

$$\langle \psi_\alpha^0 | H^0 | \psi^1 \rangle + \langle \psi_\alpha^0 | H' | \psi^0 \rangle = E^0 \langle \psi_\alpha^0 | \psi^1 \rangle + E^1 \langle \psi_\alpha^0 | \psi^0 \rangle \tag{6.34}$$

Since H^0 is Hermitian, the first term on the left cancels the first term on the right, and we get:

$$\langle \psi_\alpha^0 | H' | \psi^0 \rangle = E^1 \langle \psi_\alpha^0 | \psi^0 \rangle \tag{6.35}$$

By substituting Equation (6.29), we get:

$$a \langle \psi_\alpha^0 | H' | \psi_\alpha^0 \rangle + b \langle \psi_\alpha^0 | H' | \psi_\beta^0 \rangle = a E^1 \tag{6.36}$$

$$a W_{\alpha\alpha} + b W_{\alpha\beta} = a E^1 \tag{6.37}$$

where, $W_{\alpha\alpha} = \langle \psi_\alpha^0 | H' | \psi_\alpha^0 \rangle$, and $W_{\alpha\beta} = \langle \psi_\alpha^0 | H' | \psi_\beta^0 \rangle$

Similarly, the inner product with ψ_β^0 yields:

$$a W_{\beta\alpha} + b W_{\beta\beta} = b E^1 \tag{6.38}$$

Doing a little bit of algebra to solve for b and using the previous two equations, we get:

$$b[W_{\alpha\beta} W_{\beta\alpha} - (E^1 - W_{\alpha\alpha})(E^1 - W_{\beta\beta})] = 0 \tag{6.39}$$

If $b \neq 0$, then the equation for E^1 will yield following quadratic equation:

$$(E^1)^2 - E^1(W_{\alpha\alpha} + W_{\beta\beta}) + (W_{\alpha\alpha} W_{\beta\beta} - W_{\alpha\beta} W_{\beta\alpha}) = 0 \tag{6.40}$$

Solving the previous equation will give the following roots:

$$E_{\pm}^1 = \frac{1}{2}\left[W_{\alpha\alpha} + W_{\beta\beta} \pm \sqrt{(W_{\alpha\alpha} - W_{\beta\beta})^2 + 4|W_{\alpha\beta}|^2}\right] \tag{6.41}$$

where we have used $W_{\alpha\beta} = W_{\alpha\beta}^*$. These are splitting of the energy levels of the degenerate states due to perturbation. Therefore, perturbation causes splitting of the energy levels. It separates the two energy levels of the same energy into energy levels of two different energies.

6.1.3 Stark Effect

The application of external uniform electric field as a perturbation to an atom causes splitting of the degenerate energy levels and is called the Stark effect. The perturbation due to external electric field is given as follows:

$$H' = eE.z = eEr\cos\theta \tag{6.42}$$

Consider the previous perturbation to the ground state of a hydrogen atom. The ground state has the following wave function ($n = 1, l = 0, m = 0$):

$$\Psi_{100} = R_{10}(r)Y_{00}(\theta, \phi) = \frac{1}{\sqrt{\pi a^3}}e^{-r/a} \tag{5.146}$$

According to first-order perturbation to energy:

$$E_1^1 = \langle \psi_{100}^0 | H' | \psi_{100}^0 \rangle = \frac{eE}{\pi a^3}\int e^{-(2r/a)}(r\cos\theta)r^2 \sin\theta \, dr \, d\theta \, d\phi = 0 \tag{6.43}$$

Thus, the previous perturbation to the ground state does not have first-order correction to the energy.

What about the perturbation of $n = 2$ energy level of the hydrogen atom?

To answer the question, we need to understand parity of a wave function.

6.1.3.1 Parity

Consider that the position coordinate "r" is reflected through the origin so that "r" is replaced by $-r$ or replacing $\theta \to \pi - \theta$, $\phi \to \phi + \pi$, and leaving "r" the same, so that:

$$\psi_{nlm}(r, \theta, \phi) \to \psi_{nlm}(-r, \theta, \phi) \to \psi_{nlm}(r, \pi - \theta, \phi + \pi)$$

If

$$\psi_{nlm}(r, \pi - \theta, \phi + \pi) = \psi_{nlm}(r, \theta, \phi) \tag{6.44}$$

then it has even parity, which means that the wave function has remained unchanged under the previous operations. If

$$\psi_{nlm}(r, \pi - \theta, \phi + \pi) = -\psi_{nlm}(r, \theta, \phi) \tag{6.45}$$

then the wave function has odd parity.

The energy eigenfunctions of a spherically symmetric potential has parity that depends on quantum number "l."

Thus,

$$\psi_{nlm}(r, \ \pi - \theta, \ \phi + \pi) = (-1)^l \psi_{nlm}(r, \ \theta, \ \phi) \tag{6.46}$$

or

$$\psi_{nlm}(-r, \ \theta, \ \phi) = (-1)^l \psi_{nlm}(r, \ \theta, \ \phi)$$

As was discussed, the ground state $(1, 0, 0)$ has even parity and thus has no first-order stark effect.

The first excited state $(n = 2)$ of hydrogen is a fourfold degenerate, $(2, 0, 0)$, $(2, 1, 0)$, $(2, 1, 1)$ and $(2, 1, -1)$. To find the perturbation to the energy levels, we use the matrix method to find the perturbed energies.

Just for simplification, let us use following notation:

$$(2, \ 0, \ 0) \rightarrow |0\rangle, \ (2, \ 1, \ 0) \rightarrow |1\rangle, \ (2, \ 1, \ 1) \rightarrow |2\rangle, (2, \ 1, -1) \rightarrow |3\rangle$$

Then, $\langle \psi_{200} | H' | \psi_{200} \rangle = \langle 0 | H' | 0 \rangle = W_{00}$, and there will be 16 elements of the 4×4 matrix given as follows.

The eigenvalue equation will be:

$$\begin{pmatrix} W_{00} - \lambda & \cdots & W_{03} \\ \vdots & \ddots & \vdots \\ W_{30} & \cdots & W_{33} - \lambda \end{pmatrix} = 0 \tag{6.47}$$

Since the perturbation is odd, the only matrix elements of H' that fail to vanish are those for the unperturbed states that have opposite parity.

The nonvanishing matrix elements of H',

$$\langle 2, 1, 0 | H' | 2, 0, 0 \rangle = \frac{eE}{16\pi a^4} \int \left(1 - \frac{r}{2a}\right) e^{-(2r/a)} r e^{-(2r/a)} \cos \theta (r \cos \theta) r^2 \sin \theta \, dr \, d\theta \, d\phi$$

$$= \frac{eE}{16\pi a^4} \int_0^\infty \int_{-1}^1 r^4 \left(2 - \frac{r}{a}\right) e^{-(r/a)} w^2 \, dw \, dr = -3eEa$$

The four roots of the previous matrix equation (Equation 6.47) are $0, 0, 3eEa$ and $-3eEa$, so that there are only three perturbed energy levels.

$$E_2, \ E_2, \ E_2 + 3eEa, \ E_2 - 3eEa$$

The energy level $n = 2$ of a hydrogen atom splits into three distinguished energy levels under the influence of the external electric field.

PROBLEMS

6.1   A 1-dimensional harmonic oscillator is perturbed by an extra potential energy ax^3, where "a" is some constant. Calculate the change in energy to first- and second-order perturbation correction.

6.2   A 3-dimensional infinite cubical well undergoes the perturbation. The unperturbed potential of the well can be described as:

$$V(x, \ y, \ z) = \begin{cases} 0, & \text{if } 0 < x < L, \quad 0 < y < L, \quad 0 < z < L \\ \infty, & \text{otherwise} \end{cases}$$

The perturbation can be described as:

$$H' = \begin{cases} V_0, & \text{if } 0 < y < L/2, \quad 0 < z < L/2 \\ \infty, & \text{otherwise} \end{cases}$$

Find the first-order correction to the ground state energy.

6.3 Find all the elements of the matrix in Equation (6.47) and show that even-even and odd-odd parity matrix elements are zero and odd-even parity-only matrix element are non-zero. Determine the three energy eigenvalues by solving the characteristic equation for the matrix.

6.4 Consider a 2-dimensional harmonic oscillator. The Hamiltonian is:

$$H_0 = \frac{p_x^2}{2m} + \frac{p_y^2}{2m} + \frac{m\omega^2}{2}(x^2 + y^2)$$

Find the energy of the ground state. Are there any degenerate energy levels?

A perturbation is applied to the previous system:

$$V = bm\omega^2 xy$$

Find the first-order correction to the ground state. Does it cause splitting of the degenerate energy levels? Find their energies.

6.2 THE VARIATIONAL PRINCIPLE

The perturbation method discussed in the previous section works well if we already know the exact solution of the Schrodinger equation for an idealized system. However, in some situations, we do not know any of the exact solutions. For such systems, we use the variational method. This method is very useful in approximating the ground state energy of a system when exact solutions are not known.

According to this method, based on the understanding of the properties of a system, make a guess of the ground state wave function $|\tilde{\psi}_0\rangle$. Such a function is a trial wave function. Then, find the expectation value of the Hamiltonian using the trial wave function. This expectation value of the ground state energy has an upper bound as given by the variational principle that follows:

$$E_0 \leq \frac{\langle \tilde{\psi}_0 | H | \tilde{\psi}_0 \rangle}{\langle \tilde{\psi}_0 | \tilde{\psi}_0 \rangle} \tag{6.48}$$

Proof: Let us consider a system. H is its Hamiltonian, and $|\tilde{\psi}_0\rangle$ is the trial wave function. Also, let $|\psi_k\rangle$ be the exact wave functions of the system such that:

$$1 = \sum_{k=0}^{\infty} |\psi_k\rangle\langle\psi_k| \tag{6.49}$$

Using previous equation, we can write the trial wave function as:

$$|\tilde{\psi}_0\rangle = \sum_{k=0}^{\infty} |\psi_k\rangle\langle\psi_k|\tilde{\psi}_0\rangle \tag{6.50}$$

Also,

$$H|\psi_k\rangle = E_k|\psi_k\rangle \tag{6.51}$$

Thus,

$$H|\tilde{\psi}_0\rangle = \sum_{k=0}^{\infty} E_k |\psi_k\rangle \langle \psi_k |\tilde{\psi}_0\rangle \tag{6.52}$$

and,

$$\frac{\langle \tilde{\psi}_0 |H|\tilde{\psi}_0\rangle}{\langle \tilde{\psi}_0 |\tilde{\psi}_0\rangle} = \frac{\sum_{k=0}^{\infty} E_k \langle \tilde{\psi}_0 |\psi_k\rangle \langle \psi_k |\tilde{\psi}_0\rangle}{\sum_{k=0}^{\infty} \langle \tilde{\psi}_0 |\psi_k\rangle \langle \psi_k |\tilde{\psi}_0\rangle} \tag{6.53}$$

but, $E_k = E_k - E_0 + E_0$, and substituting it in the previous equation gives:

$$\frac{\langle \tilde{\psi}_0 |H|\tilde{\psi}_0\rangle}{\langle \tilde{\psi}_0 |\tilde{\psi}_0\rangle} = \frac{\sum_{k=0}^{\infty} (E_k - E_0)|\langle \tilde{\psi}_0 |\psi_k\rangle|^2}{\sum_{k=0}^{\infty} |\langle \tilde{\psi}_0 |\psi_k\rangle|^2} + E_0 \tag{6.54}$$

Hence,

$$\frac{\langle \tilde{\psi}_0 |H|\tilde{\psi}_0\rangle}{\langle \tilde{\psi}_0 |\tilde{\psi}_0\rangle} \geq E_0 \tag{6.55}$$

Using any trial wave function, the ground state energy of the system can be obtained. This ground state energy is always close to the exact ground state energy if the right trial wave function is used for the evaluation. Therefore, guessing the right wave function is key to using this method.

EXAMPLE 6.2

Consider a particle in an infinite square well. Find the ground state energy of the particle using the variational method. The potential has the following form:

$$V(x) = \begin{cases} 0, & -L \leq x \leq L \quad \text{(Region II)} \\ \infty, & \text{otherwise} \quad \text{(Region I and III)} \end{cases} \tag{6.56}$$

For the previous system, we know that the exact ground energy has following value:

$$E_0 = \frac{\pi^2 \hbar^2}{8 L^2 m} \tag{6.57}$$

Therefore, it will be easy to check the application of the variational method for the previous system.

As we can see from the previous form of the potential, the wave function for the previous system must vanish at $x = \pm L$. Therefore, we can guess the trial wave function to have the following form:

$$\tilde{\psi}_0 = A (L^2 - x^2) \tag{6.58}$$

As a first step, let us normalize the previous function:

$$\langle \tilde{\psi}_0 |\tilde{\psi}_0\rangle = 1 = A^2 \left[\int_{-L}^{L} (L^2 - x^2)^2 \, dx \right] \tag{6.59}$$

$$A = \sqrt{\frac{15}{16 L^5}} \tag{6.60}$$

$$\tilde{\psi}_0 = \sqrt{\frac{15}{16 L^5}} (L^2 - x^2) \tag{6.61}$$

Using the previous normalized function, the expectation value of the Hamiltonian is:

$$\langle \tilde{\psi}_0 | H | \tilde{\psi}_0 \rangle = \frac{(-\hbar^2/2m)A^2 \int_{-L}^{L} (L^2 - x^2)(d^2/dx^2)(L^2 - x^2)}{\langle \tilde{\psi}_0 | \tilde{\psi}_0 \rangle = 1} \tag{6.62}$$

$$\langle \tilde{\psi}_0 | H | \tilde{\psi}_0 \rangle = \frac{10}{\pi^2} \left(\frac{\pi^2 \hbar^2}{8 L^2 m} \right) \approx 1.013 E_0 \tag{6.63}$$

Thus, the trial wave function gives the ground state energy very close to the exact value. Finding these trial wave functions that can provide a good approximation to the ground state is the most difficult thing, but with practice, guessing such functions can be made better.

6.2.1 THE GROUND STATE OF A HELIUM ATOM

One of the very first applications of the variational method is the determination of the ground state energy of the helium atom. The helium atom is a little more complicated than the hydrogen atom because it consists of multiple electrons and is considered a three-body system. The nucleus (consisting of two protons and one or two neutrons) of the helium atom is considered a single stationary body and has two electrons orbiting around it as the other two nonstationary bodies. The nucleus is much heavier in the case of helium, and, therefore, its motion is almost negligible and can be considered a stationary body located at the center of the mass of the system. The figure that follows describes the structure of helium (Figure 6.1).

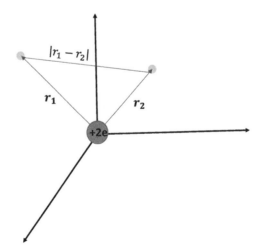

FIGURE 6.1 An illustration of a helium atom.

The Hamiltonian for this system is:

$$H = -\frac{\hbar^2}{2m}(\nabla_1^2 + \nabla_2^2) - \frac{e^2}{4\pi\varepsilon_0}\left(\frac{2}{r_1} + \frac{2}{r_2} - \frac{1}{|r_1 - r_2|} \right) \tag{6.64}$$

The first two terms are the kinetic energies of the electrons, and third term comprises the interaction potential between the two protons of the nucleus, with the two electrons (electrostatic attraction), and the interaction between the two electrons (electrostatic repulsion).

If we treat two electrons to be independent that do not influence each other, then the trial wave function is simply the product of two hydrogen atoms ground state wave function:

$$\psi_1(r_1,\ r_2) = \frac{8}{\pi a^3} e^{-2\,(r_1+r_2)/a} \tag{6.65}$$

where the factor of 8 and 2 in the previous equation is mainly due to nuclear charge $Z = 2$. If we consider that the interaction between the electrons influences their overall interaction with the nucleus, then we need to consider that each electron shields the other electron from the nucleus in such a way that each electron sees the nuclear charge much less than $Z = 2$. Thus, the trial wave function can be written as:

$$\psi_1(r_1,\ r_2) = \frac{Z^3}{\pi a^3} e^{-Z(r_1+r_2)/a} \tag{6.66}$$

Here, Z is an unknown parameter, which we need to find using the variational method. The figure that follows describes the shielding effect (Figure 6.2).

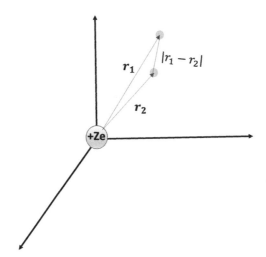

FIGURE 6.2   An illustration of a helium atom of a shielded nuclear charge $+Ze$ and two surrounding electrons.

We need to rewrite the previous Hamiltonian in terms of the parameter Z as:

$$H = -\frac{\hbar^2}{2m}\left(\nabla_1^2 + \nabla_2^2\right) - \frac{e^2}{4\pi\varepsilon_0}\left(\frac{Z}{r_1} + \frac{Z}{r_2}\right) + \frac{e^2}{4\pi\varepsilon_0}\left(\frac{Z-2}{r_1} + \frac{Z-2}{r_2} + \frac{1}{|r_1 - r_2|}\right) \tag{6.67}$$

The ground state energy will be the expectation value of the previous Hamiltonian using the trial wave function described in Equation (6.66).

The first three terms are simply twice the total energy of the hydrogen atom in the ground state, and using this understanding, the expectation value of Hamiltonian is:

$$\langle\psi_1|H|\psi_1\rangle = 2\,Z^2 E_1 + (Z-2)\left(\frac{e^2}{4\pi\varepsilon_0}\right)\left\langle\psi_1\left|\frac{1}{r_1} + \frac{1}{r_2}\right|\psi_1\right\rangle + \frac{e^2}{4\pi\varepsilon_0}\left\langle\psi_1\left|\frac{1}{|r_1 - r_2|}\right|\psi_1\right\rangle \tag{6.68}$$

$$\left\langle\psi_1\left|\frac{1}{r_1}\right|\psi_1\right\rangle = \left\langle\psi_1\left|\frac{1}{r_2}\right|\psi_1\right\rangle = \left\langle\psi_1\left|\frac{1}{r}\right|\psi_1\right\rangle = \left(\frac{Z^3}{\pi a^3}\right)^2\int e^{-2Zr'/a}\,d^3r'\int e^{-2Z\,r/a}\left(\frac{1}{r}\right)d^3r \tag{6.69}$$

Since, $\dfrac{Z^3}{\pi a^3} \displaystyle\int e^{-2Zr'/a} d^3 r' = 1$

$$\langle \psi_1 | \tfrac{1}{r} | \psi_1 \rangle = \frac{Z^3}{\pi a^3} \int_0^\infty e^{-2Z\,r/a} \left(\frac{1}{r} \right) r^2\, dr \int_0^\pi \sin\theta\ d\theta \int_0^{2\pi} d\varphi = \frac{4Z^3}{a^3} \int_0^\infty e^{-2Z\,r/a} r\, dr \qquad (6.70)$$

$$\frac{4Z^3}{a^3} \int_0^\infty e^{-2Z\,r/a}\ r\, dr = \frac{4Z^3}{a^3} \left\{ \left(r\, e^{-2Zr/a} \right)_0^\infty + \left(\frac{a}{2z} \right) \int_0^\infty e^{-2Zr/a}\, dr \right\}$$

$$= \frac{4Z^3}{a^3} \left(\frac{a^2}{4\, Z^2} \right)$$

$$= \frac{Z}{a}$$

Thus,

$$\left\langle \psi_1 \left| \frac{1}{r} \right| \psi_1 \right\rangle = \frac{Z}{a} \qquad (6.71)$$

and,

$$\langle \psi_1 | H | \psi_1 \rangle = 2\, Z^2 E_1 + 2(Z - 2) \left(\frac{e^2}{4\pi\varepsilon_0} \right) \left(\frac{Z}{a} \right) + \frac{e^2}{4\pi\varepsilon_0} \left\langle \psi_1 \left| \frac{1}{|r_1 - r_2|} \right| \psi_1 \right\rangle \qquad (6.72)$$

Now, we need to evaluate the last term in the previous equation:

$$\left\langle \psi_1 \left| \frac{1}{|r_1 - r_2|} \right| \psi_1 \right\rangle = I_1 = \left(\frac{Z^3}{\pi a^3} \right)^2 \int e^{-2Zr_1/a}\, d^3 r_1 \int \left(\frac{1}{r_{12}} \right) e^{-2Z\, r_2/a}\, d^3 r_2 \qquad (6.73)$$

$$r_{12} = |r_1 - r_2| = \sqrt{r_1^2 + r_2^2 - 2r_1 r_2 \cos(\gamma)}$$

where γ is the angle between r_1 and r_2 vectors. The angle between the vectors can be obtained from their scalar product.

$$\vec{r_1} \cdot \vec{r_2} = r_1 r_2 (\cos\theta_1 \cos\theta_2 + \sin\theta_1 \sin\theta_2 \cos\theta_1 \cos(\varphi_2 - \varphi_1)) \qquad (6.74)$$

$\theta_1, \varphi_1, \theta_2, \varphi_2$ are the polar angles of the vectors r_1 and r_2.
Thus,

$$\cos(\gamma) = \cos\theta_1 \cos\theta_2 + \sin\theta_1 \sin\theta_2 \cos\theta_1 \cos(\varphi_2 - \varphi_1) \qquad (6.75)$$

The function $1/r_{12}$ can be expanded in terms of spherical harmonics.

$$\frac{1}{r_{12}} = \frac{1}{r_1} \sum_{n=0}^\infty \left(\frac{r_2}{r_1} \right)^n P_n(\cos\gamma) \quad r_1 > r_2$$

$$= \frac{1}{r_2} \sum_{n=0}^\infty \left(\frac{r_1}{r_2} \right)^n P_n(\cos\gamma) \quad r_2 > r_1 \qquad (6.76)$$

When we substitute Equation (6.76) into Equation (6.73) and use the orthogonality of spherical harmonics, the integration over the polar angles of r_1 causes all terms to vanish except that for $n = 0$.

Thus, the integral takes the following form:

$$I_1 = \left(\frac{Z^3}{\pi a^3}\right)^2 \int_0^\infty \left[(4\pi)^2 \left(\frac{1}{r_1}\int_0^{r_1} r_2^2\, e^{-2Z(r_2/a)}\, dr_2 + \int_{r_1}^\infty r_2\, e^{-2Z(r_2/a)}\, dr_2\right)\right] r_1^2\, e^{-2Zr_1/a}\, dr_1 \qquad (6.77)$$

$$4\pi \left(\frac{1}{r_1}\int_0^{r_1} r_2^2\, e^{-2Z(r_2/a)}\, dr_2 + \int_{r_1}^\infty r_2\, e^{-2Z(r_2/a)}\, dr_2\right) = \frac{4\pi a^3}{4Z^3 r_1}\left[1 - \left(1 + \frac{Zr_1}{a}\right) e^{-2Z(r_1/a)}\right] \qquad (6.78)$$

Thus,

$$I_1 = 4\pi \left(\frac{Z^3}{\pi a^3}\right)^2 \frac{4\pi a^3}{4Z^3}\int_0^\infty \frac{1}{r_1}\left[1 - \left(1 + \frac{Zr_1}{a}\right) e^{-2Z(r_1/a)}\right] e^{-2Z\, r_1/a}\, r_1^2 dr_1$$

$$= 4\pi \left(\frac{Z^3}{\pi a^3}\right)\int_0^\infty r_1\left[1 - \left(1 + \frac{Zr_1}{a}\right) e^{-2Z(r_1/a)}\right] e^{-2Z\, r_1/a}\, dr_1 \qquad (6.79)$$

Finally,

$$\left\langle \psi_1 \left| \frac{1}{|r_1 - r_2|} \right| \psi_1 \right\rangle = \frac{5Z}{8a}$$

and

$$\frac{e^2}{4\pi\varepsilon_0}\left\langle \psi_1 \left| \frac{1}{|r_1 - r_2|} \right| \psi_1 \right\rangle = \frac{e^2}{4\pi\varepsilon_0}\left(\frac{5Z}{8\,a}\right) = -\frac{5Z}{4}E_1 \qquad (6.80)$$

where E_1 is the ground state energy of a hydrogen atom.
Collecting all the terms obtained previously:

$$\langle \psi_1 | H | \psi_1 \rangle = 2\, Z^2 E_1 - 4Z\,(Z - 2)E_1 - \frac{5Z}{4}E_1 = \left[-2Z^2 + \frac{27}{4}\, Z\right]E_1 \qquad (6.81)$$

But we still do not know the value of Z. Let us find it.

$$\frac{d\langle H\rangle}{dZ} = 0 = \left[-4Z + \frac{27}{4}\right] = 0 \qquad (6.82)$$

$$Z = \frac{27}{16} = 1.69$$

Substituting the value of Z in Equation (6.81), we get:

$$\langle H\rangle = \frac{27^2 E_1}{128} = -77.5\,\text{eV} \qquad (6.83)$$

The experimental value of the ground state of helium atom was found to be -78.975 eV. Therefore, the theoretical value shown here is quite close to the experimental value. This example shows that the variational method provides a good approximation to the ground state energy of a helium atom. The result that for $Z = 27/16$ rather than $Z = 2$, the variational method gives the ground state energy close to the experimental value, indicating that each electron screens the nucleus from the other charge, thus reducing the effective nuclear charge by 5/16 (Equation 6.68) of the electronic charge.

Conceptual Question 2: *Explain the scenarios in which each of the approximation methods is better over the other.*

Conceptual Question 3: *Consider that an experimentalist has determined the ground state energy of his system but is not sure whether the value he has is correct or not, and he asks you (a theoretical physicist) for your help to find whether his measured value is right or not. Which method (perturbation theory or variational method) would you use to validate his answer? Why?*

PROBLEMS

6.5 Use the following trial wave function, $\psi(x) = Ae^{-\alpha x^2}$ to find the ground state energy of a 1-dimensional harmonic oscillator where A is a normalization constant, and α is some constant that you need to determine.

6.6 Use the following trial wave function $\psi(r) = Ae^{-r/a}$ to find the ground state energy of a particle of mass "m" bound by a potential $V(r) = V_0 e^{-r/a}$. Here, V_0 is some constant.

6.7 Using the Hamiltonian and wave function in Equations (6.64) and (6.65), determine the ground state energy of a helium atom. Compare this value with the experimental value.

6.8 A particle of mass "m" is under the influence of a potential $V(x) = \alpha x^4$. Construct a trial wave function based on the properties that at $x = 0$ the function will be maximum so that V is minimum, has even parity, and vanishes at $x \rightarrow \infty$. Using this function, find the ground state energy of the particle.

6.3 THE WKB METHOD

The Wentzel-Kramers-Brillouin (WKB) method is mainly used for approximating the wave function of quantum systems interacting with time-independent, slowly varying potentials. It is called semi-classical approximation because slowly varying potentials do not give rise to quantum mechanical features arising from the wave-particle duality of matter. In such cases, the classical description is adequate to explain the phenomenon. It is mostly useful in the nearly classical limit of large quantum numbers.

The time-independent Schrodinger equation for a 1-dimensional quantum system can be written as:

$$-\frac{\hbar^2}{2m}\frac{d^2\psi(x)}{dx^2} + V(x)\psi = E\psi$$

or

$$\frac{d^2\psi(x)}{dx^2} + \frac{2m}{\hbar^2}(E - V(x))\psi = 0 \tag{6.84}$$

Classically, the total energy of such a system is given as:

$$\frac{p(x)^2}{2m} + V(x) = E \tag{6.85}$$

where $p(x)$ is the momentum of the system or particle which is a function of space, $V(x)$ is the potential energy function, and E is the total energy. This implies that the momentum (or kinetic energy) of the particle varies at every point in space. Using the de Broglie wave-particle duality relationship, we can also write the wavelength of the particle as a function of space:

$$\lambda(x) = \frac{h}{p(x)} \tag{6.86}$$

where $p(x) = \sqrt{2m(E - V(x))} > 0$; otherwise, $p(x)$ is imaginary. The previous expression (Equation 6.86) is neither classical nor quantum mechanical (momentum p is not an operator). It is purely semi-classical. We will use Equation (6.85) to write the Schrodinger equation as:

$$\left[\frac{d^2}{dx^2} + \frac{p^2}{\hbar^2}\right]\psi(x) = 0 \tag{6.87}$$

Let us write the wave function as some sinusoidal function. As we discussed in the previous chapters, the wave function of most 1-dimensional quantum systems can be approximated as a sinusoidal function. Therefore,

$$\psi(x) = A(x)\,e^{i\phi(x)} \tag{6.88}$$

where $A(x)$ is the amplitude of the wave function which is a function of space, and $\phi(x)$ is the phase. By substituting the previous function into Equation (6.87), we get:

$$\frac{d^2A(x)}{dx^2} + 2i\frac{dA(x)}{dx}\frac{d\phi(x)}{dx} + iA(x)\frac{d^2\phi(x)}{dx^2} - A(x)\left(\frac{d\phi(x)}{dx}\right)^2 = -\frac{p^2}{\hbar^2}A(x) \tag{6.89}$$

The real and imaginary parts of the previous equation can be separated into two equations as:

$$\frac{d^2A(x)}{dx^2} - A(x)\left(\frac{d\phi(x)}{dx}\right)^2 = -\frac{p^2}{\hbar^2}A(x) \tag{6.90}$$

$$2i\frac{dA(x)}{dx}\frac{d\phi(x)}{dx} + iA(x)\frac{d^2\phi(x)}{dx^2} = 0 \tag{6.91}$$

The slowly varying potential ensures that the amplitude of the wave function also varies slowly and by very small amount. Thus, the second derivative of the amplitude in Equation (6.90) can be neglected. This is called the WKB approximation. The equation then becomes:

$$\left(\frac{d\phi(x)}{dx}\right)^2 = \frac{p^2}{\hbar^2} \tag{6.92}$$

Taking the integral of the previous equation, we get:

$$\phi(x) = \pm\frac{1}{\hbar}\int p(x)\,dx \tag{6.93}$$

Thus, the phase of the wave function has physical meaning, as its rate of change with position is equal to the momentum of the particle.

Equation (6.91) can be simply written as:

$$\frac{d}{dx}\left[A^2 \frac{d\phi(x)}{dx}\right] = 0 \tag{6.94}$$

Thus,

$$A^2 \frac{d\phi(x)}{dx} = \text{const} = C \tag{6.95}$$

where C is some real number

$$A(x) = \sqrt{\frac{\hbar C}{p(x)}} = \sqrt{\frac{A(x_0)^2 p(x_0)}{p(x)}} \tag{6.96}$$

where the term on the right-hand side can be obtained from the initial conditions.

Now, we can write the wave function as:

$$\psi(x) = A(x)e^{\pm (i/\hbar) \int p(x)dx} \tag{6.97}$$

Using Equation (6.96), the previous equation can be written as:

$$\psi(x) = A(x_0)\sqrt{\frac{p(x_0)}{p(x)}} \exp\left[\pm \frac{i}{\hbar}\int_{x_0}^{x} p(x)\,dx\right] \tag{6.98}$$

If we define, $\psi(x_0) = A(x_0)$, then we can write the previous function as:

$$\psi(x) = \psi(x_0)\sqrt{\frac{p(x_0)}{p(x)}} \exp\left[\pm \frac{i}{\hbar}\int_{x_0}^{x} p(x)\,dx\right] \tag{6.99}$$

We can see that the probability density of the particle in space at some location 'x' is:

$$P(x) = |\psi(x)|^2 \alpha \frac{1}{p(x)} = \frac{1/m}{v(x)} \tag{6.100}$$

Thus, the probability of finding the particle at a certain location in space is inversely proportional to the classical velocity of the particle. It will be hard to locate the particle if it is moving fast. This is exactly what is to be expected in a classical ensemble because the time spent by a particle is inversely proportional to the velocity in that region. Therefore, $P(x)$ is the same as the classical distribution function.

Now we can see why the amplitude of the wave function must vary very slowly as given here:

$$|\psi(x)| \alpha \frac{1}{\sqrt{p(x)}} \alpha \frac{1}{\sqrt{v(x)}} \tag{6.101}$$

Thus, the WKB approximation will be a good approximation where $V(x)$ is a sufficiently smooth and slowly varying function.

EXAMPLE 6.3

Consider the problem of a particle in a potential well, whose potential function $V(x)$ is defined as (Figure 6.3)

$$V(x) = \begin{cases} F(x), & 0 < x < L \\ \infty, & \text{otherwise} \end{cases} \tag{6.102}$$

Now using the WKB method, let us write the phase of the wave function, which is:

$$\phi(x) = \pm \frac{1}{\hbar} \int p(x)\, dx \tag{6.103}$$

Inside the well, $p(x)$ is real since $E > V(x)$ and the wave function can be written as:

$$\psi(x) = \frac{1}{\sqrt{p(x)}} \{ B_+ e^{i\phi(x)} + B_- e^{-i\phi(x)} \} \tag{6.104}$$

which can be further simplified in the following form:

$$\psi(x) = \frac{1}{\sqrt{p(x)}} \{ B_1 \sin\phi(x) + B_2 \cos\phi(x) \} \tag{6.105}$$

Within the well, the phase can be written as:

$$\phi(L) - \phi(0) = \frac{1}{\hbar} \int_0^L p(x)\, dx \tag{6.106}$$

The boundaries of the infinite well are also called turning points. The WKB approximation very near these points must be still valid, and wave function is finite. However, the potential energy

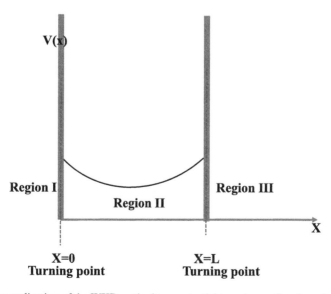

FIGURE 6.3 The application of the WKB method to a potential trough; $x = 0$ and $x = L$ are turning points.

becomes infinite at these points; therefore, the wave function must vanish at these points:

$$\psi(x = 0) = 0, \quad \text{and} \quad \phi(0) = 0, \quad B_2 = 0$$

$$\psi(x = L) = 0, \quad \text{and} \quad \phi(L) = n\pi, \quad B_2 = 0$$

where $n = 1, 2, 3\ldots$
   Thus,

$$\int_0^L p(x)\, dx = n\pi\hbar \tag{6.107}$$

This is a quantization condition for the momentum and allowed energies that we have obtained directly by solving the Schrodinger equation in Chapter 4.

$$\int_0^L \sqrt{2m(E - V(x))}\, dx = n\pi\hbar \tag{6.108}$$

If $V(x) = 0$ within the well, then we have:

$$\sqrt{2mE_n}L = n\pi\hbar$$

Thus,

$$E_n = \frac{n^2\pi^2\hbar^2}{2mL^2} \tag{6.109}$$

is the quantized energy of the particle inside the well.

EXAMPLE 6.4 TUNNELING THROUGH A BARRIER

By the early nineteenth century, it was observed experimentally that alpha particle (α), which comprises of two protons and two neutrons, was spontaneously emitted from the nucleus of a radioactive atom. According to classical physics, such a particle cannot be emitted from the nucleus due to the strong coulombic force of the nucleus that will keep the alpha particle bounded to the nucleus. The experimental observation of these particles baffled physicists at that time.

   The explanation of this phenomenon of alpha particle decay was one of the earliest applications of quantum mechanics. The explanation was provided by George Gamow and independently by Condon and Gurney in 1928. It was the very first application of quantum mechanics to nuclear physics and was in very good agreement with the experimental results (Figure 6.4).

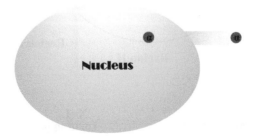

FIGURE 6.4   The schematic illustrates tunneling of an alpha particle through a nucleus of the atom.

According to Gamow's model, within the nucleus, the potential can be considered as the spherically symmetric constant potential $V(r)$ of the following form ($V(r) = -V_0$). Outside the nucleus, the potential is $V(r) = +Z\,Z'e^2/r$, where Z is the atomic nuclear charge of the final nucleus (daughter nucleus), and $Z' = +2$ is the nuclear charge. This is the nuclear repulsion between the daughter nucleus and the particle in region II.

The previous potential has the following form (Figure 6.5)

$$V(r) = \begin{cases} -V_0, & 0 < r < R \\ \dfrac{ZZ'e^2}{r}, & R < r < \infty \end{cases} \tag{6.110}$$

Using the WKB method, the wave function of the α-particle is:

$$\psi(r) = \frac{B}{\sqrt{|p(r)|}}\left[\mathrm{Exp}\left[\pm\frac{1}{\hbar}\int |p(r)|\,dr\right]\right] \tag{6.111}$$

Here $p(r)$ is imaginary since the energy of the particle is much less than the of the coulombic potential

where:

$$p(r) = \left[2\mu\left[V(r) + \frac{\hbar^2 l(l+1)}{2\mu r^2} - E\right]\right]^{1/2} \tag{6.112}$$

centrifugal term is included in the previous expression. The probability of transmission through the barrier is:

$$P = \exp\left[-\frac{2}{\hbar}\int_{r_1}^{r_2}\left\{2\mu\left[V(r) + \frac{\hbar^2 l(l+1)}{2\mu r^2} - E\right]\right\}^{1/2}dr\right] \tag{6.113}$$

and,

$$P = \exp[-2\gamma] \tag{6.114}$$

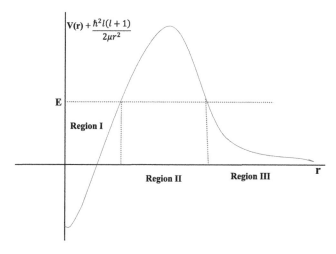

FIGURE 6.5 The application of the WKB method to a potential barrier, with radial coordinate; the centrifugal term is included.

The expression shows that the probability of penetrating through the potential barrier of the nucleus is an exponentially decaying function. Lifetime for α-particle decay is expected to be of the order of some characteristic time τ, divided by the function P obtained from Equation (6.107).

In all previous applications of the WKB method, we have not discussed the boundary conditions where the potential function changes drastically at the turning points, where classical regions do not transition to nonclassical regions very smoothly, and our approximation breaks down. For such situations, we need to join the WKB solution from the two regions. In the section that follows, we discuss such scenarios.

6.3.1 Turning Points of a Bound State

So far, we have not discussed in detail the difficulties that arise for obtaining solutions at the boundaries using the WKB method. Consider a particle interacting with a potential function as shown in Figure 6.6.

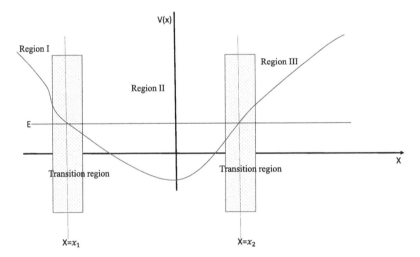

FIGURE 6.6 The WKB approximation to the wave function works well except in the shaded regions near the classical turning points x_1 and x_2. After approximating the potential by a linear function within each region, the Schrodinger equation is solved in these regions, and then wave functions ψ_I, ψ_{II}, ψ_{III} are joined.

The turning points x_1 and x_2 are the classical turning points for the energy E. In regions I and III, the energy of the particle is much less than the potential energy, momentum is imaginary, and wave function is exponentially decaying function. Classically, such a region is called forbidden region. In region III, the wave function can be written as:

$$\psi_{III}(x) \sim \frac{1}{\sqrt{2m[V(x) - E]}}\left[\text{Exp}\left[-\frac{1}{\hbar}\int \sqrt{2m[V(x') - E]}\, dx'\right]\right] \tag{6.115}$$

In region II, which is a classically allowed region, the wave function will be a periodic function shown as:

$$\psi_{II}(x) = \frac{B}{\sqrt{p(x)}}\left[\sin\left[\frac{1}{\hbar}\int p(x')\, dx' + C\right]\right] \tag{6.116}$$

where B, C are some real constants

However, at the turning point, for instance at $x = x_1$, $V(x_1) = E$, and thus both of the previous functions blow up. At $x = x_1$, $p(x_1) \to 0$, and wave length tends to infinity, and the WKB approximation that the potential function varies slowly at this point requires that the wave length of the particle also varies slowly fails. In other words, the WKB approximation breaks down, and the approximate wave functions obtained by solving the Schrodinger equation in each region cannot be matched at the turning points. How do we solve such a problem?

To solve this problem, we need to define the region near the turning points as some transition region where we need to solve the Schrodinger equation. Since there are two turnings points, there are two transition regions. In these regions, if the potential is assumed to be a slowly varying function, then it must be some linear function as:

$$
V(x) \sim V(x_1) + \left.\frac{dV}{dx}\right|^{x=x_1} (x - x_1)
$$

$$
= E + V'(x - x_1)
$$

$$(6.117)$$

The exact solution of the Schrodinger equation using the previous potential in these regions is obtained and is then matched. For simplicity, we will avoid showing here the mathematical steps leading to solutions of the Schrodinger equations in these regions. Rather, we will discuss the final solutions. In region II, near the turning point x_1, the solution of the Schrodinger equation gives the following wave function:

$$
\psi_{II}(x) = \frac{B}{\sqrt{p(x)}} \left[\sin\left[\frac{1}{\hbar} \int_{x_1}^{x} p(x') \, dx' - \frac{\pi}{4}\right] \right]
$$

$$(6.118)$$

Near turning point x_2, the wave function is:

$$
\psi_{II}(x) = \frac{B'}{\sqrt{p(x)}} \left[\sin\left[\frac{1}{\hbar} \int_{x_2}^{x} p(x') \, dx' + \frac{\pi}{4}\right] \right]
$$

$$(6.119)$$

For the previous two solutions to match, $B = B'$, and the difference in the phase of the two sine functions must be a multiple of π:

$$
\frac{1}{\hbar} \int_{x_1}^{x} p(x') \, dx' - \frac{\pi}{4} - \frac{1}{\hbar} \int_{x_2}^{x} p(x') \, dx' - \frac{\pi}{4} = n\pi
$$

$$(6.120)$$

Thus,

$$
\int_{x_1}^{x_2} p(x) \, dx = \left(n + \frac{1}{2}\right) \pi \hbar
$$

$$(6.121)$$

$n = 0, 1, 2, 3, \ldots$

The previous equation is called the energy quantization rule. Using this rule, the quantized energies of a quantum system can be determined. This approximation allows us to find the quantized energies without solving the Schrodinger equation.

EXAMPLE 6.5

Consider a particle under the influence of a linear potential, $V(x) = k|x|$. Find the quantized energy levels of the particle.

We need to use the turning points to evaluate the integral in Equation (6.122) to determine energy levels.

Turning points are, $x_1 = E/k$, $x_2 = -E/k$. Thus, the integral is:

$$\int_{x_1}^{x_2} p(x)\, dx = \int_{-E/k}^{E/k} \sqrt{2m(E - V(x))}\, dx = \left(n + \frac{1}{2}\right)\pi\hbar \qquad (6.122)$$

$$\int_{-E/k}^{E/k} \sqrt{2m(E - V(x))}\, dx = 2\int_{0}^{E/k} \sqrt{2m(E - kx)}\, dx \qquad (6.123)$$

To simplify the integral, let use define:

$$z = \frac{k}{E} x \qquad (6.124)$$

$$2\int_{0}^{1} \frac{E}{k} \sqrt{2mE(1 - z)}\, dz = \frac{2}{k}.2^{3/2}E^{3/2}m^{1/2} = \left(n + \frac{1}{2}\right)\pi\hbar \qquad (6.125)$$

Thus,

$$E_n = \left(\frac{k(n + (1/2))\pi\hbar}{2^{5/2}m^{1/2}}\right)^{2/3} \qquad (6.126)$$

Conceptual Question 4: Why is the WKB called as a semiclassical approximation? Explain.

Conceptual Question 5: Plot the linear potential function discussed in Example 6.5, and describe the turning points.

PROBLEMS

6.9   Derive Equation (6.96) using Equation (6.95) and Equation (6.94).

6.10   For the potential well (Example 6.3), show that the probability density of the particle is not inversely proportional to its velocity.

6.11   Find the allowed energy levels of a harmonic oscillator ($V(x) = \frac{1}{2}m\omega^2 x^2$) using the WKB method.

6.12   Consider a particle under the influence of a potential $V(x) = k|x|^3$. Find the quantized energy levels of the particle using WKB approximation.

6.13   Explain clearly what is semiclassical about the WKB method. Is the momentum of a particle considered as quantum mechanical if it is a function of space?

7 Time-Dependent Perturbation Theory

7.1 INTRODUCTION

In all the applications of the formalism of quantum theory, we have considered mainly time-independent potential functions, $V(r, t) = V(r)$. Such potentials made it easier to solve exactly the Schrodinger's equation or by solving the Schrodinger's equation approximately by using perturbation theory. The solution of the Schrodinger equation for time-independent potentials is obtained by using method of separable variables. The solution of time-independent Schrodinger gives stationary states, and the time dependence of the wave function is obtained by adding a phase factor to the stationary state. The schematic that follows describes such an approach (Figure 7.1).

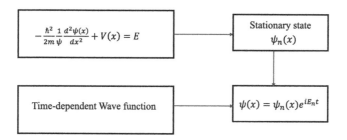

FIGURE 7.1 A schematic illustrating the method of separation of variables.

However, such an approach works only for time-independent potentials. When a quantum system's potential is time dependent, the approach of separation of variables cannot be used. In this chapter, we will develop approaches for solving the Schrödinger's equation with time-dependent potential functions. First, we will begin with understanding time-dependent perturbation theory. According to this method, when the time-dependent potential is small, then we can treat it as a perturbation to our system. The Hamiltonian can be written as:

$$H = H_0 + H'(t) \tag{7.1}$$

The time-dependent part of the Hamiltonian causes transitions from one stationary state to the another state of the unperturbed Hamiltonian, H_0. Such transitions are also called quantum jumps between one energy level and another, or between one stationary state and another. Due to such transitions, the coefficients of superposition state of a wave function of the system changes with time in a more complex way. In this approach, our goal is to determine such coefficients. Using such coefficients, we can determine probabilities for such transitions. To determine such coefficients, let us follow the approach of the time-independent perturbation theory.

Conceptual Question 1: The time-dependent perturbation theory is applicable for certain kinds of potentials only. Describe those potentials by giving examples.

7.2 FIRST-ORDER PERTURBATION THEORY

To develop this approach, we consider the state vector instead of the wave function. This simplifies obtaining transition probabilities. Consider the Schrodinger equation in terms of a state vector or state ket:

$$ih\frac{\partial|\Psi(t)\rangle}{\partial t} = \hat{H}|\Psi(t)\rangle \tag{7.2}$$

Here, H is the Hamiltonian as given in Equation (7.1). Our goal is to solve the previous equation. Since the perturbation $H'(t)$ is small, we can still write the state ket in terms of stationary states or eigenkets of H_0 ($|E_n^0\rangle$).

$$|\psi(t)\rangle = \sum_n c_n(t)|E_n^0\rangle \tag{7.3}$$

In Chapter 3, we learned that the state of a particle in terms of its spin states for the time-independent Hamiltonian can be written as:

$$|\Psi,t\rangle = c_\uparrow(t)|\uparrow\rangle + c_\downarrow(t)|\downarrow\rangle \tag{7.4}$$

where

$$c_\uparrow(t) = c_\uparrow(0)e^{-\frac{i\omega t}{2}} \tag{7.5}$$

$$c_\downarrow(t) = c_\downarrow(0)e^{\frac{i\omega t}{2}} \tag{7.6}$$

Similarly, for the time-dependent Hamiltonian, we write the state vector as:

$$|\psi(t)\rangle = \sum_n a_n(t)e^{-iE_n^0 t/\hbar}|E_n^0\rangle \tag{7.7}$$

where

$$c_n(t) = a_n(t)e^{-iE_n^0 t/\hbar} \tag{7.8}$$

The coefficient a_n changes with time because of the time-dependent perturbation. By substituting Equation (7.7) into Equation (7.2), we get:

$$0 = \sum_n \left[i\hbar\frac{da_n}{dt} - H'(t)a_n(t)\right]e^{-iE_n^0 t/\hbar}|E_n^0\rangle \tag{7.9}$$

In the previous equation, there is no dependence on H_0, and it completely depends on the perturbation part. By multiplying the previous equation by $\langle E_m^0|e^{iE_m^0 t/\hbar} >$, we get:

$$i\hbar\frac{da_m}{dt} = \sum_n \left[\langle E_m^0|H'(t)|E_n^0\rangle\right]a_n(t)\, e^{i(E_m^0 - E_n^0)t/\hbar} \tag{7.10}$$

In the previous equation, we can define the Bohr frequency as:

$$\omega_{mn} = \frac{E_m^0 - E_n^0}{\hbar} \tag{7.11}$$

Substituting the previous expression into Equation (7.10), we obtain:

$$i\hbar\frac{da_m}{dt} = \sum_n \left[\langle E_m^0|H'(t)|E_n^0\rangle\right]a_n(t)\, e^{i\omega_{mn}t} \tag{7.12}$$

In arriving at the previous equation, we completely eliminated the time-independent part of the Hamiltonian. To apply perturbation theory, let us replace H' by $\lambda H'$ in Equation (7.1) and replace the expansion coefficients by the power series:

$$a_n = a_n^0 + \lambda a_n^1 + \lambda^2 a_n^2 + \cdots \tag{7.13}$$

Here, λ varies between 0 and 1. By substituting Equation (7.13) and $\lambda H'$ in Equation (7.12), we obtain the following equations:

$$\frac{da_m^0}{dt} = 0 \tag{7.14}$$

$$i\hbar \frac{da_m^1}{dt} = \sum_n [\langle E_m^0 | H'(t) | E_n^0 \rangle] \, a_n^0(t) \, e^{i\omega_{mn} t} \tag{7.15}$$

Equation (7.14) shows that zeroth order coefficients a_m^0 are constants in time. Their values are the initial conditions of the problem and specify the state of the system before the perturbation is applied. Initially, the system is in an unperturbed state, a_m^0 can be assumed to be:

$$a_n^0 = \langle E_n^0 | E_k^0 \rangle = \delta_{nk} \tag{7.16}$$

By substituting Equation (7.16) into Equation (7.15) we get:

$$i\hbar \frac{da_m^1}{dt} = \sum_n [\langle E_m^0 | H'(t) | E_k^0 \rangle] \, e^{i\omega_{nk} t} \tag{7.17}$$

The previous equation can be simply integrated to obtain the coefficient as:

$$a_m^1(t) = \frac{1}{i\hbar} \int_0^t \langle E_m^0 | H'(t') | E_k^0 \rangle e^{i\omega_{mk} t'} \, dt' \tag{7.18}$$

To first-order correction, the expansion coefficient of the wave function can be described as:

$$a_n(t) = \delta_{nk} + \frac{1}{i\hbar} \int_0^t \langle E_n^0 | H'(t') | E_k^0 \rangle e^{i\omega_{nk} t'} \, dt' \tag{7.19}$$

Consider a two-level system, whose energy eigenvectors are, $|0\rangle$, $|1\rangle$ corresponding to energies E_0 and E_1. Some "time" dependent perturbation $H'(t)$ is applied to the system between time 0 to t. Determine the transition probabilities from state $|0\rangle$ to $|1\rangle$, when the particle transitions from the lower state and from $|1\rangle$ to $|0\rangle$ when the particle transitions from the higher state.

Since the system is initially in the lower state, therefore, zeroth order perturbation of the higher state must be zero. Therefore,

$$a_1^0 = 0 \tag{7.20}$$

The expansion coefficient for up to first-order correction is:

$$a_1(t) = \frac{1}{i\hbar} \int_0^t \langle E_1^0 | H'(t') | E_0^0 \rangle e^{i\omega_{10} t'} \, dt' \tag{7.21}$$

The transition probability is:

$$P_{0 \to 1} = |a_1(t)|^2 \tag{7.22}$$

EXAMPLE 7.1

Let us consider that the previous time-dependent perturbation is of the following form:

$$H'(t) = \gamma \, e^{-i\omega t} |E_1^0\rangle\langle E_0^0| \tag{7.23}$$

γ is some real constant, and ω is driving frequency of the perturbation. Calculate the transition probability from lower to higher state.

By substituting Equation (7.23) into Equation (7.19), we get:

$$a_1(t) = \frac{\gamma}{i\hbar} \int_0^t e^{-i\omega t'} e^{i\omega_{10} t'} \, dt' \tag{7.24}$$

$$a_1(t) = \frac{\gamma}{i\hbar} \left[\frac{e^{i(\omega_{10}-\omega)t} - 1}{i(\omega_{10} - \omega)} \right] \tag{7.25}$$

$$a_1(t) = -\frac{\gamma}{\hbar} e^{-i\frac{(\omega_{10}-\omega)}{2} t} \left[\frac{e^{i\frac{(\omega_{10}-\omega)}{2} t} - e^{-i\frac{(\omega_{10}-\omega)}{2} t}}{(\omega_{10} - \omega)} \right] \tag{7.26}$$

$$a_1(t) = -2i\frac{\gamma}{\hbar} e^{-i\frac{(\omega_{10}-\omega)}{2} t} \left[\frac{\sin \frac{(\omega_{10} - \omega)}{2} t}{(\omega_{10} - \omega)} \right] \tag{7.27}$$

Thus, the transition probability is:

$$P_{0\to1} = |a_1(t)|^2 = 4\left(\frac{\gamma}{\hbar}\right)^2 \left[\frac{\sin \frac{(\omega_{10} - \omega)}{2} t}{(\omega_{10} - \omega)} \right]^2 \tag{7.28}$$

The transition probability depends on the frequency difference between the external perturbation and the energy difference between the levels, $\omega_{10} - \omega$. For $\omega_{10} - \omega = \pi$, the transition probability is maximum.

Conceptual Question 2: *What is the effect of a small time-dependent perturbation on a system? Explain it by giving examples.*

EXAMPLE 7.2

Consider a 1-dimensional harmonic oscillator in the ground state $|0\rangle$ at time $t = -\infty$, and some time-dependent perturbation. Equation (7.29) is applied at time "t" between $t = -\infty$ to $t = \infty$. Find the transition probability to state $|n\rangle$ at $t = \infty$.

$$H'(t) = -e \in \left(\frac{\hbar}{2m\omega}\right)^{\frac{1}{2}} (a + a^\dagger) e^{-t^2/\tau^2} \tag{7.29}$$

The expansion coefficient for the wave function of a perturbed system can be written as:

$$a_n(t = \infty) = \frac{1}{i\hbar} \int_{-\infty}^{\infty} \langle n|H'(t')|0\rangle e^{i\omega_{10} t'} \, dt' \tag{7.30}$$

The previous integral will be nonvanishing only for $n = 1$. Thus, the transition will be from level 0 to 1.

$$a_1(t = \infty) = \frac{1}{i\hbar} \int_{-\infty}^{\infty} \langle 1|H'(t')|0\rangle e^{i\omega_{10}t'} dt'$$

$$= \frac{ie \in}{\hbar} \left(\frac{\hbar}{2m\omega}\right)^{\frac{1}{2}} \int_{-\infty}^{\infty} e^{-t^2/\tau^2} e^{i\omega_{10}t} dt \qquad (7.31)$$

$$= \frac{ie \in}{\sqrt{2\hbar m\omega}} (\pi\tau^2)^{1/2} e^{-\omega^2\tau^2/4}$$

The transition probability is:

$$P_{0\to 1} = |a_1(t = \infty)|^2 = \frac{e^2 \in^2 \pi\tau^2}{2\hbar m\omega} e^{-\omega^2\tau^2/2} \qquad (7.32)$$

Conceptual Question 3: *Consider a particle in an infinite square well that is in a nth energy level (E_n). Compare the quantum behavior of the particle with and without time-dependent perturbation.*

7.3 PERIODIC PERTURBATIONS

Consider that a periodic perturbation is applied to a system Equation (7.33). By periodic, it means that perturbation is maximum and minimum at different times, and thus, the strength of the perturbation varies with time. How does the periodic nature of perturbation affect the transition probability for transition from initial state "i" to a final state "n"?

$$H'(r, t) = \hat{V}_{ni} \cos(\omega t) \qquad (7.33)$$

where $\hat{V}_{ni} = V_0|n\rangle\langle i|$ and V_0 is some real number. The expansion coefficient is:

$$a_n(t) = \frac{1}{i\hbar} \int_0^t \langle n|H'(t')|i\rangle e^{i\omega_{ni}t'} dt' = \frac{V_0}{2i\hbar} \int_0^t (e^{i\omega t'} + e^{-i\omega t'}) e^{i\omega_{ni}t'} dt' \qquad (7.34)$$

$$= \frac{-V_0}{2\hbar} \left[\frac{e^{i(\omega_{ni}+\omega)t} - 1}{\omega_{ni} + \omega} + \frac{e^{i(\omega_{ni}-\omega)t} - 1}{\omega_{ni} - \omega}\right] \qquad (7.35)$$

where ω_{ni} is the transition frequency, and ω is the driving frequency of the perturbation.

If the driving frequency and the transition frequencies are close, then first term in the previous equation approaches zero (because the denominator is very large), and the only term left is the second. Thus, the expression is:

$$a_n(t) = \frac{-V_0}{2\hbar} \left[\frac{e^{i(\omega_{ni}-\omega)t} - 1}{\omega_{ni} - \omega}\right] = \frac{-V_0}{2\hbar} e^{i(\omega_{ni}-\omega)t/2} \left[\frac{e^{i(\omega_{ni}-\omega)t/2} - e^{-i(\omega_{ni}-\omega)t/2}}{\omega_{ni} - \omega}\right] \qquad (7.36)$$

$$a_n(t) = \frac{-V_0}{2\hbar} \left[\frac{e^{i(\omega_{ni}-\omega)t} - 1}{\omega_{ni} - \omega}\right] = \frac{iV_0}{\hbar} e^{i(\omega_{ni}-\omega)t/2} \left[\frac{\sin(\omega_{ni} - \omega)t/2}{(\omega_{ni} - \omega)}\right] \qquad (7.37)$$

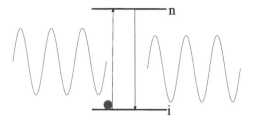

FIGURE 7.2 This figure illustrates sinusoidal transition probability. If the perturbation is turned on for a long time, then the particle keeps transitioning back and forth between lower and excited states.

The transition probability is:

$$P_{i \to n} = |a_n(t)|^2 = \frac{V_0^2}{\hbar^2} \left[\frac{\sin(\omega_{ni} - \omega)t/2}{(\omega_{ni} - \omega)} \right]^2 \tag{7.38}$$

The transition probability is periodic and oscillates back and forth between maximum and minimum values. The maximum transition probability is $\frac{V_0^2}{\hbar^2} \left[\frac{1}{(\omega_{ni} - \omega)} \right]^2$ and minimum is zero. It means that at certain times the chances of the particle to transition to an excited state is high. At times when the transition probability is zero, the particle returns to a lower state or stays in an initial state. If the perturbation is turned on for a long time, then the particle keeps transitioning back and forth between lower and excited states, but if the main purpose of the perturbation is to allow the particle to transition to an excited state, then perturbation cannot be applied for a long time. For the particle to transition to an excited state permanently, the perturbation should be turned off at a time when the transition probability is maximum. Figure 7.2 describes the transition probabilities.

PROBLEMS

7.1 Calculate the transition probability for a system when external perturbation is constant in time. Do this problem by replacing the time-dependent perturbation in Example 7.2 by a constant potential.

7.2 Using Equation (7.5), derive the expansion coefficients for a two-level system when external time-dependent perturbation is applied.

7.3 A hydrogen atom is in the ground state at $t = -\infty$. An electric field $E(t) = E_0 \hat{k} e^{t^2/\tau^2}$ is applied until $t = \infty$. Calculate the transition probability of the atom to any excited state.

7.4 THE SUDDEN APPROXIMATION

In all previous discussions related to perturbation theory, we discussed the change in the Hamiltonian of the system due to perturbation leading to a change in the wave function. Such perturbation causes transitions between the states. However, in some situations, the perturbation can be for such a short time, almost "instantaneous," that the change in the Hamiltonian does not result in a change in the wave function.

To understand it more clearly, consider a system whose Hamiltonian changes suddenly over a small interval of time τ such that $\tau \to 0$. The Schrodinger equation of the system is:

$$i\hbar \frac{\partial |\Psi(t)\rangle}{\partial t} = \hat{H}(t)|\Psi(t)\rangle \tag{7.39}$$

Integrate the previous equation over time interval, $-\tau/2$ to $\tau/2$, so that the time interval over which sudden perturbation is applied is still τ:

$$|\Psi(t = \tau/2)\rangle - \left|\Psi\left(t = -\frac{\tau}{2}\right)\right\rangle = \frac{1}{i\hbar} \int_{-\tau/2}^{\tau/2} H(t)|\Psi(t)\rangle dt \qquad (7.40)$$

In the limit $\tau \to 0$, the integral on the right vanishes, therefore,

$$|\Psi(t = -\tau/2)\rangle = \left|\Psi\left(t = \frac{\tau}{2}\right)\right\rangle \qquad (7.41)$$

which means that the wave function remains undisturbed if there is instantaneous change in the Hamiltonian.

What about the transition probability? Will the system jump to a lower or an excited state as result of sudden perturbation?

According to Equation (7.41), the wave function of the system will remain same. Therefore, if the system is initially in an eigenstate or stationary state, such a state will remain undisturbed as a result of instantaneous change in the Hamiltonian. Thus, sudden perturbation will not cause any transition or jumps between the states. For this reason, transition probability will be zero.

Let us apply above reasoning to Example 7.2. In this example, a 1-dimensional harmonic oscillator undergoes perturbation for some time τ. Because of perturbation, the transition probability for transition from ground state to first excited state is:

$$P_{0 \to 1} = |a_1(t = \infty)|^2 = \frac{e^2 \mathcal{E}^2 \pi \tau^2}{2\hbar m \omega} e^{-\omega^2 \tau^2/2}$$

Thus, transition probability depends on duration perturbation, which is τ. Now, if $\tau \to 0$ then

$$P_{0 \to 1} = 0$$

But saying that $\tau \to 0$ may not seem like a realistic situation. In a real scenario, duration will be small but not zero. In such a realistic scenario, will the transition probability be still zero?

To answer such a question, let us consider the Hamiltonian of a system that suddenly changes from H_0 to H_1 at time $t = 0$ and remains same thereafter. Because of perturbation, the new energy eigenfunctions of the system is:

$$H_1 v_m = E_m v_m \qquad (7.42)$$

Before the perturbation is applied, the eigenfunctions were, $u_m e^{-iE_m t/\hbar}$ and, let say that system was in some initial eigenstate, but after the perturbation, the solution of the Schrodinger equation is the eigenfunction shown in Equation (7.42). However, the solution of the Schrodinger equation at $t = 0$ must connect with the solution at a later time "t"; thus, the wave function can be written as:

$$\psi(x) = u_n(x) = \sum_m C_{mn} v_m(x) \qquad (7.43)$$

where coefficients C_{mn} can be obtained by using orthonormality condition. In this way, the wave function of the system does not change before and after the perturbation. The wave function is now a linear superposition state of the eigenfunctions of the changed Hamiltonian. It is no more an eigenfunction or stationary state of the previous Hamiltonian. Since the wave function is in a superposition state, it can transition to any eigenstate state of the changed Hamiltonian.

To understand the previous findings, consider the case of β decay from an atom of nuclear charge Z. An atom of charge Z undergoes β-decay by emitting a relativistic electron and changing its charge to $Z + 1$. The time the emitted electron takes to get out of the first shell ($n = 1$) is:

$$\tau \cong a_0/Zc \tag{7.44}$$

$a_0 =$ some constant, and $c =$ the velocity of light.

The characteristic time, which is the time electron spends in the first shell, is:

$$T = \frac{a_0}{Z^2 \alpha c} \tag{7.45}$$

Thus,

$$\frac{\tau}{T} = Z\alpha \tag{7.46}$$

For Z to be very small, we may say that we applied the perturbation for a very small time, and such perturbation did not change the wave function of the electron. However, such a wave function is now a linear superposition of the eigenstate of the new atom of charge ($Z + 1$) (Equation 7.43). This wave function no longer corresponds to the stationary state of the new atom of charge ($Z + 1$). This implies that there is a certain probability that the atom will be left in an excited state of the new atom because of the sudden process of β-decay. This excitation can be observed by the detection of an emitted radiation by the atom. In the previous sections, we showed that time-dependent perturbation causes the transitions to the other energy levels of the unperturbed Hamiltonian. In this way, sudden perturbation causes the changes equivalent to that of time-dependent perturbation.

Conceptual Question 4: *Consider a particle in an infinite square well that is in a ground state. A sudden perturbation is applied to the well to double its size. Will this perturbation cause transition to the ground state of the new well? Explain why or why not.*

PROBLEMS

7.4 Consider the wave function given in Equation (7.43). Use the orthonormality condition for the eigenfunctions to obtain a mathematical relation for the expansion coefficients C_{mn}.

7.5 Consider that a sudden perturbation is applied to a harmonic oscillator that was initially in a ground state. The perturbation applies a constant force ($F = $ const) for a time period that is much smaller than the period of oscillation of the harmonic oscillator. Determine the probability that the atom is found in first excited state after the force has been turned off.

7.5 ADIABATIC APPROXIMATION

In the previous sections on perturbation theory, we learned that a perturbation to a system causes changes to the energy, wave function of the system, and causes transitions between the energy states. For example, if a system was initially in a stationary state before perturbation was applied to the system, then the effect of the perturbation would be to transition it to some other stationary state, to change the stationary state to a linear superposition state and so on. Thus, perturbation causes changes to a system, which means it causes changes to the parameters of a system such as stationary state, energy, momentum and position. The changes caused by perturbation to a system make it difficult to predict the behavior of the system. What if there are perturbations that do not cause changes to certain parameters of the system so that certain parameters of the system remain unchanged as a result of the perturbation?

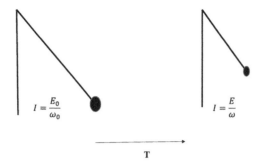

FIGURE 7.3 An illustration of the adiabatic invariant in a pendulum.

7.5.1 ADIABATIC PROCESS

In classical physics, adiabatic processes are mainly associated with thermodynamics and mechanics. In thermodynamics, when changes occur to a system such that it remains in a state of equilibrium with its environment at all the times, such that its entropy remain constant, the process by which such changes occur to a system is called an adiabatic process. For example, consider a gas inside a container that is a thermal insulator. If now the gas expands or contracts without gaining or losing heat energy, then such a process of expansion or contraction of gas is called an adiabatic process. Heat energy and entropy remain constant all the time; thus, heat energy and entropy are called adiabatic invariants.

In mechanical systems, if one of the parameters defining the motion of a system is varied very slowly over a long period of time, the energy is no longer constant in time. However, the action variable for such a system remains constant in time and, thus, is an adiabatic invariant. For example, consider a pendulum whose length "l_0" is reduced to some length "l" very slowly over a long period of time (T). The action variable of the pendulum which is $I = \frac{E}{\omega}$, (where E is the energy and ω is the angular frequency of the pendulum) remains constant in time. After a very long time, "T," the energy and angular frequency of oscillations of the pendulum changes, but action variable which is adiabatic invariant remains constant, as shown in Figure 7.3. The shortening of the string increases the energy and angular frequency of the pendulum.

In quantum mechanics, consider a Hamiltonian of a system "$H(0)$" at time $t = 0$ which changes very slowly over a very long period of time "T" to $H(T)$. If initially the system starts out in a stationary state $|n(0)\rangle$, what would be the final state of the system after time "T"?

To find an answer to such a question, we need to resort to adiabatic theorem.

7.5.2 ADIABATIC THEOREM

The adiabatic theorem was conceived by Albert Einstein and Paul Ehrenfest in 1911 and later, proved by Max Born and Vladimir A. Fok in 1927.

According to this theorem, if the Hamiltonian of a system changes very slowly with time, then the solution of Schrodinger's equation at any instant of time can be approximated by means of a stationary eigenfunction of the instantaneous Hamiltonian, so that a particular eigenfunction at one time goes over continuously into corresponding eigenfunction at a later time.

Consider a quantum mechanical system that starts in a stationary state $|n(0)\rangle$ at time $t = 0$. According to this theorem, if the system changes slowly with time, then after time "T" the system will be found in stationary state $|n(t)\rangle$ of the system at that time. The adiabatic invariant of the system is the quantum number "n." For example, consider a particle in the ground state of an infinite square well initially, and the well expands slowly with time. After, time "T," the particle will be in the ground state of the expanded well. Thus, slow changes to the length of the well causes the particle to transition to the ground state of the well at time "T," and quantum number $n = 1$ (ground state)

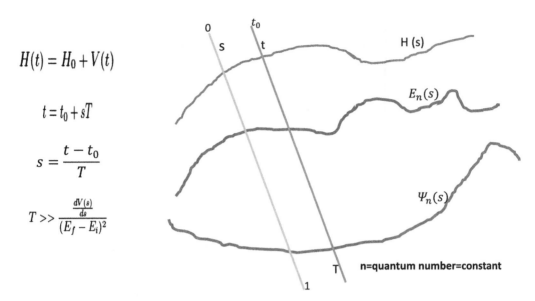

$$H(t) = H_0 + V(t)$$

$$t = t_0 + sT$$

$$s = \frac{t - t_0}{T}$$

$$T \gg \frac{\frac{dV(s)}{ds}}{(E_f - E_i)^2}$$

FIGURE 7.4 This schematic illustrates using doodles as the changing Hamiltonian, energy and wave function of a quantum mechanical system whose parameter "s" is changed adiabatically.

remains the same and is an adiabatic invariant. In classical physics, entropy and action integral are adiabatic invariant, but in quantum mechanics, the quantum number is the adiabatic invariant.

$$|n(0)\rangle \rightarrow |n(T)\rangle \tag{7.47}$$

$n = $ constant, adiabatic invariant.

For the adiabatic process, though the system interacts with its environment, its quantum number remains invariant (Figure 7.4).

For any ideal quantum system, a particle that is not interacting with its environment, we expect the particle that was in some initial stationary state at $t = 0$ to remain in the same stationary state after time "$t = T$" but acquires a dynamical phase factor. Thus, the stationary state of the particle after time "T" is following:

$$|n(T)\rangle = e^{-iE_nT/\hbar}|n(0)\rangle \tag{7.48}$$

For such an ideal system, except for the phase factor, energy and quantum number remain the same. The phase factor ($e^{-iE_nT/\hbar}$), is also called the dynamic phase.

For a quantum system that is undergoing adiabatic changes means that it is interacting with the environment, its energy is changing with time, but its quantum number remains the same. It continuously transitions from one stationary state to another stationary state of the same quantum number. So how different is the stationary state after time "t" of a quantum system that has changed adiabatically than that of an ideal system, Equation (7.48)?

This question was answered by M.V. Berry in 1983. According to him, the stationary state acquires an additional phase along with the dynamic phase due to adiabatic changes in the system. Thus, for a quantum system changing adiabatically, the stationary state after time "t" will have the following form:

$$\psi(t) = \exp\left\{-\frac{i}{\hbar}\int_0^t E_n(t')dt'\right\}e^{i\gamma_n t}|n(t)\rangle \tag{7.49}$$

The first term is the usual dynamic phase factor, and the second phase factor $(e^{i\gamma_n t})$ is called Berry's phase or the geometric phase because the phase depends on the geometric properties of the system.

7.5.3 PROOF OF THE ADIABATIC THEOREM

To derive the previous equation, let us consider a quantum system that is evolving adiabatically with time. After time "t," its state vector will acquire a phase factor, thus

$$\psi(t) = e^{i\phi(t)}|n(t)\rangle \tag{7.50}$$

is the state of the system after time "t." The Schrodinger equation for the system is:

$$i\hbar\frac{d\psi(t)}{dt} = H(t)\psi(t) \tag{7.51}$$

where $H(t)$ is the Hamiltonian of the system that is varying very slowly with time. By substituting Equation (7.50) into Equation (7.51), the following is obtained:

$$\frac{d\phi}{dt} = -\frac{1}{\hbar}\langle\psi(t)|H(t)|\psi(t)\rangle + i\langle n(t)|\frac{d}{dt}|n(t)\rangle \tag{7.52}$$

By integrating the previous equation,

$$\phi(t) = -\frac{1}{\hbar}\int_0^T E_n(t)dt + i\int_0^T \langle n(t)|\frac{d}{dt}|n(t)\rangle \, dt$$

The previous phase comprises of two terms. The first term

$$\theta = -\frac{1}{\hbar}\int_0^T E_n(t)\,dt \tag{7.53}$$

is the usual dynamical phase, and the second term is:

$$\gamma_n(t) = i\int_0^T \langle n(t)|\frac{d}{dt}|n(t)\rangle \, dt \tag{7.54}$$

and called Berry's phase or the geometric phase. This phase does not directly depend on the Hamiltonian.

To further understand the meaning of Equation (7.54), consider that $R(t)$ is a parameter of a quantum system that is changing very slowly with time, so that the system is changing adiabatically, and the stationary state of the system after time "t" will be $|n(R(t))\rangle$. The geometric phase will depend on the parameter $R(t)$ and is given as:

$$\gamma_n(t) = i\int_{R_0}^{R(t)} \langle n(R(t'))|\nabla_R n(R(t'))\rangle dR(t') \tag{7.55}$$

Thus, the geometric phase depends on the parameter $R(t)$ of the system that has changed slowly with time. It only depends on the final and initial values of the parameter $R(t)$, not on the path. This clearly shows that by controlling the parameter $R(t)$ externally, we can make the quantum state to acquire a required geometric phase.

The previous equation is simplified if the parameter $R(t)$ describes a closed loop, so that after a certain period of time, it comes back to the same point where it started, $R(t=T) = R(t=0)$, where T is the time period after which the parameter returns to the same value. Using Stokes' theorem for transforming the line integral to a surface integral, the previous equation is:

$$\gamma_n(T) = i \oint \langle n(R)|\nabla_R n(R)\rangle dR = - \iint \chi_n(R).dS \tag{7.56}$$

$$\chi_n(R) = Im \, \nabla_R \, X \langle \, n(R)|\nabla_R n(R)\rangle \tag{7.57}$$

The curl of two functions has the following property:

$$\nabla X[f(x)\nabla g(x)] = \nabla f(x) X \, \nabla g(x) \tag{7.58}$$

Using the previous property of the curl, Equation (7.52) is:

$$\chi_n(R) = Im \, (\nabla_R \langle n(R)|) \, X \, (\nabla_R |n(R)\rangle) \tag{7.59}$$

$$= Im \sum_{m \neq n} [\nabla_R \langle n(R)|m(R)\rangle] \, X[\langle m(R)|\nabla_R|n(R)\rangle] \tag{7.60}$$

The term with $m = n$ is purely imaginary and, therefore, is omitted from the previous equation. The Hamiltonian of such a system is also parameter $R(t)$ dependent:

$$H(t) \rightarrow H(R(t)) \tag{7.61}$$

$$H(R(t))|n(R(t))\rangle = E_n(R(t))|n(R(t))\rangle \tag{7.62}$$

For simplicity,

$$R(t) \rightarrow R \tag{7.63}$$

and taking the gradient of the previous equation,

$$\nabla_R[H(R)|n(R)\rangle = E_n(R)|n(R)\rangle] \tag{7.64}$$

$$\langle m(R)|\nabla_R H(R)|n(R)\rangle + \langle m(R)|H(R)\nabla_R|n(R)\rangle = \nabla_R E_n(R) \, \langle m(R)|n(R)\rangle + E_n \langle m(R)|\nabla_R|n(R)\rangle \tag{7.65}$$

$$\langle m(R)|\nabla_R H(R)|n(R)\rangle + E_m(R)\langle m(R)|\nabla_R|n(R)\rangle = E_n(R)\langle m(R)|\nabla_R|n(R)\rangle \tag{7.66}$$

$$\langle m(R)|\nabla_R|n(R)\rangle = \frac{\langle m(R)|\nabla_R H(R)|n(R)\rangle}{E_n - E_m} \quad m \neq n \tag{7.67}$$

By substituting the previous equation in Equation (7.60), the following is obtained:

$$\chi_n(R) = Im \sum_{m \neq n} \frac{\langle n(R)|\nabla_R H(R)|m(R)\rangle X \langle m(R)|\nabla_R H(R)|n(R)\rangle}{(E_n - E_m)^2} \tag{7.68}$$

The geometrical phase after one complete period is:

$$\gamma_n(T) = i \oint \langle n(R)|\nabla_R n(R)\rangle dR = - \iint \chi_n(R).dS \tag{7.69}$$

where χ_n is given by Equation (7.68).

Conceptual Question 5: *In thermodynamics, heat and entropy are adiabatic invariants, and in classical mechanics, "action" is an adiabatic invariant. What is a quantum mechanical adiabatic invariant? Explain. Give an example.*

EXAMPLE 7.3

Consider an electron that is interacting with a magnetic field. The direction of the magnetic field is changed slowly with time, but the magnitude remains constant. The magnetic field direction is changed in such a way that it completes one loop as shown in figure in time "*T*." The Hamiltonian of such a system is:

$$\hat{H}(B(t)) = -g\mu\,\hat{S}.B(t) \tag{7.70}$$

Here, magnetic field is the parameter R that is changed slowly, and $\hat{S}$ is the spin of the electron. For simplicity, the magnetic field is along the z-axis of the spin space such that:

$$H(R(t))|n(R(t))\rangle = E_n(R(t))|n(R(t))\rangle$$

Thus,

$$H(B(t))|\uparrow\rangle = -g\mu\,\hat{S}_z B(t)|\uparrow\rangle = \frac{-g\mu\hbar B(t)}{2}|\uparrow\rangle \tag{7.71}$$

$$\nabla_B\hat{H}(B) = -g\mu\hat{S} \tag{7.72}$$

$$\hat{S} = \hat{S}_x\hat{i} + \hat{S}_y\hat{j} + \hat{S}_z\hat{k} \tag{7.73}$$

$$\hat{S}_z = \frac{\hbar}{2}[|\uparrow\rangle\langle\uparrow| - |\downarrow\rangle\langle\downarrow|] \tag{7.74}$$

$$\hat{S}_x = \frac{\hbar}{2}[|\uparrow\rangle\langle\downarrow| + |\downarrow\rangle\langle\uparrow|] \tag{7.75}$$

$$\hat{S}_y = \frac{i\hbar}{2}[-|\uparrow\rangle\langle\downarrow| + |\downarrow\rangle\langle\uparrow|] \tag{7.76}$$

Using Equation (7.68), $\chi_n(B)$ can be evaluated, the first term

$$\langle n(R)|\nabla_R\hat{H}(R)|m(R)\rangle = (-g\mu)\langle\uparrow|\hat{S}|\downarrow\rangle \tag{7.77}$$

$$\chi_n(B) = (-g\mu)^2\,\frac{\left[\dfrac{\hbar}{2}\hat{i} - \dfrac{i\hbar}{2}\hat{j}\right] X \left[\dfrac{\hbar}{2}\hat{i} + \dfrac{i\hbar}{2}\hat{j}\right]}{(-g\mu B)^2\left(\dfrac{\hbar}{2} + \dfrac{\hbar}{2}\right)^2} \tag{7.78}$$

$$\chi_n(B_x) = \chi_n(B_y) = 0, \quad \chi_n(B_z) = \frac{\hat{k}}{2B^2} \tag{7.79}$$

The geometric phase is given as:

$$\gamma_n(T) = -\int\int \chi_n(R).dS = -\int\int \frac{\hat{k}}{2B^2} dB = -\frac{1}{2}\int d\Omega = -\frac{\Omega}{2} \tag{7.80}$$

where Ω is the solid angle subtended at the origin by the path taken by the magnetic field. This is independent of the magnitude of the magnetic field and energy of the electron, unlike the dynamical phase. It simply depends on the quantum number $(m = 1/2)$ corresponding to the spin of the electron.

PROBLEMS

7.6   For the spin "m" of a particle, such that $S_z = m\hbar(m = -S, +S)$, that is interacting with very slowly varying magnetic field whose magnitude is constant but direction slowly changes with time, determine the geometric phase of the particle as previously discussed.

7.7   Consider an electron interacting with a time-dependent magnetic field,

$$B(t) = B_0(\sin\theta\cos\omega t\,\hat{i} + \sin\theta\sin\omega t\,\hat{j} + \cos\theta\,\hat{k})$$

where the magnitude of the magnetic field is constant. The precession of the electron along with the magnetic field make it to sweep out a cone in space, of opening angle θ with frequency ω. Find the state vector of the electron after time "t" and the transition probabilities? Compare the behavior of the electron with above electron (adiabatic approximation). Explain the differences between the behavior of the electron for these two situations?

7.8   Consider a particle in an infinite square well, whose wall is expanding adiabatically with time $(L(t) = a + bt)$. If initially the particle started out in the ground energy state, what would be it energy state at time "t"? Find the dynamical phase factor.

7.6   MEASUREMENT PROBLEM REVISITED

In Chapter 2 we discussed the problem of measurement in quantum mechanics. We learned that, according to the Copenhagen interpretation, it is the measurement process that causes a collapse of the wave function, and the act of measurement brings out a wave or a particle aspect of a quantum mechanical system. This interpretation was not acceptable to Albert Einstein. In 1935, he and his collaborators published a paper that suggested that quantum mechanical reality is incomplete. In the section that follows, we describe in detail the measurement problem.

Consider the decay of the neutral pi meson (pion) into an electron and a positron:

$$\pi^0 = e^- + e^+ \tag{7.81}$$

The pion had spin zero before decay; therefore, after the decay, the electron and the positron must have their spins add up to zero due to the conservation of angular momentum. Thus, the electron and the positron pair must be in spin singlet state, shown as:

$$|\text{Spin singlet}\rangle = \frac{1}{\sqrt{2}}\left(|\uparrow_1\downarrow_2\rangle_x - |\downarrow_1\uparrow_2\rangle_x\right) \tag{7.82}$$

Here, x-direction is chosen as the axis of quantization. For simplicity, let us call "electron" as particle 1 in spin-up state and "positron" as particle 2 in spin-down state.

According to quantum mechanics, before any measurement is performed, particle 1 and particle 2 are in spin-up and spin-down states simultaneously.

However, the two particles are strongly correlated, and such correlation is called entanglement between the particles. This entanglement means that:

$$|\text{Spin singlet}\rangle = \frac{1}{\sqrt{2}} (|\uparrow_1 \downarrow_2\rangle_x - |\downarrow_1 \uparrow_2\rangle_x) \neq \frac{1}{\sqrt{2}}(|\uparrow_1\rangle_x + |\downarrow_1\rangle_x)\frac{1}{\sqrt{2}}(|\downarrow_2\rangle_x - |\uparrow_2\rangle_x) \quad (7.83)$$

If a measurement of the spin is performed on particle 1, then there is a 50% chance that it will be found in spin-up state or spin-down state. However, if measurement outcome of particle 1 yields a spin-up state, then the measurement outcome of particle 2 will yield a spin-down state with certainty and vice versa.

For example, consider a beam of particles where each particle is prepared in the spin-singlet state (entangled state) with another particle. Thus, two pairs of particles exist in a singlet state, and the beam has multiple pairs like this one. Before any measurement, the two particles of each pair are in an entangled state. If a measurement is now performed on particle 1 of one of the pairs and the outcome yields a spin-up state, then the measurement on particle 2 will yield a spin-down state with certainty. If measurement is now performed on particle 1 of the second pair, and particle 1 is found in a spin-down state, then the measurement on particle 2 will yield a spin-up state with certainty. Repeating the measurements will yield the same outcomes. This clearly shows that the measurement outcome of the second particle is dependent on the measurement outcome of the first particle.

Let us say that measurement of spin-up is registered as positive (+) by the detector, and measurement of spin-down is registered as negative (−) by the detector. Table 7.1 describes the measurement outcomes of a beam of entangled particles passing through the detectors.

This shows that there are strong correlations between the particles. Upon measurement, if particle 1 chooses to be in a spin-up state, then particle 2 will be found in a spin-down state with certainty and vice versa. One can even predict the measurement outcome of particle 2 with certainty before any measurement is performed on it. Thus, the measurement is a selection process. It is the measurement process that causes the particles to appear in a spin-up or spin-down state. If no measurement is performed, then particles are not in a well-defined state such as spin-up or spin-down state but rather are in an entangled state of the spins and are strongly correlated. The measurement process reveals such correlation between the particles. This is called the Copenhagen interpretation. In fact, these correlations are so strong that even if the particles are miles apart (separated in space), and measurement is performed on the particles, the measurement will yield the same results.

However, if no measurement is performed on particle 1, and measurement is performed on particle 2, then there is a 50% chance that the measurement outcome of particle 2 will yield a spin-up or spin-down state.

Table 7.2 describes such measurement outcomes.

Thus, the measurement outcome of particle 2 when no measurement is performed on particle 1 is random and Table 7.2 describes the measurement outcome of particle 2 when no measurement is performed on particle 1.

TABLE 7.1

The Measurement Outcomes of Two Entangled Particles

Measurement Outcome of Particle 1	Measurement Outcome of Particle 2
+	−
−	+
+	−
+	−
−	+
−	+

TABLE 7.2

The Measurement Outcome of Particle 2 When No Measurement is Performed on Particle 1

No Measurement on Particle 1	Measurement Outcome of Particle 2
	+
	+
	−
	+

Similarly, if there exists no strong correlation between the two particles, and the two particles are in a quantum state shown here:

$$\frac{1}{2}(|\uparrow_1\rangle_x + |\downarrow_1\rangle_x)(|\downarrow_2\rangle_x - |\uparrow_2\rangle_x) \tag{7.84}$$

then the outcome of the measurements on the two particles is random and uncorrelated. Also, when a pion decays into an electron and a positron, the electron and the positron cannot exist in this state because of nonconservation of angular momentum.

Before any measurement, the two particles are in a linear superposition state of spin-up and spin-down states simultaneously, but there are no correlations between them. If measurement of the spin state is now performed on particle 1 and it yields a spin-up value, then there is a 50% chance that the measurement outcome of particle 2 will yield a spin-up or spin-down value. Thus, the measurement outcome of particle 2 is random and is independent of the measurement on particle 1.

Table 7.3 describes the repeated measurement outcomes.

Thus, there are no correlations between the particles, and the measurement outcome of particle 2 is random.

TABLE 7.3

The Measurement Outcomes of Particles 1 and 2 When the Two Particles are Unentangled

Measurement Outcome of Particle 1	Measurement Outcome of Particle 2
+	+
−	+
+	−
−	−

7.6.1 EPR PARADOX

In 1935, Einstein and his colleagues, Boris Podolsky and Nathan Rosen, published a paper where they argued that the measurement outcome of particle 2 cannot depend on the measurement outcome of particle 1, and such correlations cannot be nonlocal. This is known as the EPR (Einstein, Podolsky and Rosen) paradox. Before any measurement is performed, the particles are in a well-defined spin-up or spin-down state. The measurement just reveals the spin state of the particles as they already exist in nature. Also, any correlation between the particles must be local because nothing can travel faster than the speed of light.

According to them, in quantum mechanics, if two variables (X and P) of a quantum mechanical system do not commute, then it is impossible to determine their values simultaneously. Any attempt to measure the X variable with certainty will disturb the P variable in such a way that its value cannot be measured with certainty.

Now consider that the spin angular momentum variables of a quantum mechanical system along X-axis and Y-axis do not commute. This implies that the spin angular momentum along X and Y axes cannot be determined with certainty simultaneously. Let us say that for two particles, particle 1 and particle 2, their angular momentum along the X-axis are in a singlet state, and their Y-component of angular momentum are also in a singlet state. If a measurement of the X-component of the spin angular momentum of particle 1 is measured and it is found to be in spin-up state, then with certainty, the spin angular momentum along X-axis of particle 2 is in spin-down state. Let us now perform measurement of the spin angular momentum of the Y-axis of particle 2. If the outcome of measurement of particle 2 yields a spin-up state, then the Y-component of the angular momentum of particle 1 is in a spin-down state with certainty. Thus, both X- and Y-components of particle 1 and particle 2 can be determined with certainty, but according to quantum mechanics, such components cannot be determined simultaneous with certainty.

Here is a clipping from their EPR paper (Figure 7.5). Einstein, A., Podolsky, B. & Rosen, N. 1935. Can quantum-mechanical description of reality be considered complete? *Physical Review* **47**(10), 777.

From this follows that either (1) the quantum mechanical description of reality given by the wave function is not complete or (2) when the operators corresponding .to two physical quantities do not commute the two quantifies cannot have simultaneous reality. For if both of them had simultaneous reality — and thus definite values — these values would enter into the complete description, according to the condition of completeness. If then the wave function provided such a complete description of reality, it would contain these values; these would then be predictable. This not being the case, we are left with the alternatives stated.

FIGURE 7.5 A clipping from EPR paper.

Let us now go back to the singlet state. According to their (EPR) argument, if particle 1 and particle 2 are either in spin state as $|\uparrow_1\downarrow_2\rangle_x$, or spin state as $|\downarrow_1\uparrow_2\rangle$, the outcome of measurement on the state of a particle in such states reveals the state of particle that it already exists in. Let say particles are in state, $|\uparrow_1\downarrow_2\rangle_x$ and if a measurement is performed on particle 1, then the outcome is a spin-up state with certainty, and the measurement outcome of particle 2 is a spin-down state with certainty. Now, if particles are in state, $|\downarrow_1\uparrow_2\rangle$, then the measurement outcome on particle 1 will yield a spin-down state with certainty, and the measurement outcome on particle 2 will yield a spin-up state with certainty. Thus, when a pion decays into an electron and a positron, there is a 50% chance that it will decay into $|\uparrow_1\downarrow_2\rangle$ or $|\downarrow_1\uparrow_2\rangle$. The singlet state describes both of these possibilities. Thus, if a beam of such particles is passed through the detectors, then half of them will be found in either of these states. The table that follows describes such outcomes.

The measurement outcomes of Tables 7.1 and 7.4 are identical. However, quantum mechanical interpretation is very different than the EPR argument. This lead them to conclude that a quantum mechanical description of reality is incomplete, and the dynamic behavior at an atomic level is probabilistic only because of an incomplete knowledge of the system. The measurement outcome of one of the particles cannot affect the measurement outcome of the other particle. The correlations between the particles are local.

TABLE 7.4

The Measurement Outcome of a Singlet State as Suggested by EPR Paper, without Particles Having Any Strong Correlations

Measurement Outcome of Particle 1	Measurement Outcome of Particle 2
+	−
−	+
+	−
−	+
+	−
−	+

The question is, who is right—Einstein's interpretation or the Copenhagen interpretation? It took some time before the issue was resolved. In 1964, John Stewart Bell came up with an inequality that could be verified experimentally. This was known as Bell's inequality. According to Bell, if the experiment shows a violation of the inequality, then the Copenhagen interpretation is right. The correlations between the particles are not local, unlike as suggested by Einstein.

7.6.2 BELL'S INEQUALITY

Consider that a particle has three spin angular momentum components, S_x, S_y and S_z. Let us start with three unit vectors $\hat{x}$, $\hat{y}$, and $\hat{z}$, and these vectors in general are not mutually orthogonal. Spin-up detection is represented as (+), and spin-down detection is represented as (−). Let us say that a particle has spin components as $(\hat{x}+, \hat{y}-, \hat{z}+)$. For such a particle, if S. $\hat{x}$ is measured, the detector registers (+) with certainty; if S. $\hat{y}$ is measured, the detector registers (−) with certainty; and if S. $\hat{z}$ is measured, then the detector registers (+) with certainty.

Let us say that there are local correlations between the particles of the two beams whose components of angular momentum are measured. The angular momentum of the two particles in the beams must add up to zero so that the total angular momentum is conserved. For example, in the two beams, a population N_1 of the particles has angular momentum as $(\hat{x}+, \hat{y}-, \hat{z}+)$ and $(\hat{x}-, \hat{y}+, \hat{z}-)$ such that total angular momentum of all the components adds up to zero and is conserved. Table 7.5 gives populations of the particles in the two beams with the local correlations.

Let us suppose that a measurement of S. $\hat{x}$ is performed on a particle from beam 1, and the detector registers (+), and then a measurement of S. $\hat{y}$ is performed on a particle from beam 2, and the detector registers (−). From Table 7.5, such a pair belong to either N_1 or N_2 population. Thus, the number of particles for which such a detection is registered are $(N_1 + N_2)$.

TABLE 7.5

Population	Particle 1	Particle 2
N_1	$(\hat{x}+, \hat{y}+, \hat{z}+)$	$(\hat{x}-, \hat{y}-, \hat{z}-)$
N_2	$(\hat{x}+, \hat{y}+, \hat{z}-)$	$(\hat{x}-, \hat{y}-, \hat{z}+)$
N_3	$(\hat{x}+, \hat{y}-, \hat{z}+)$	$(\hat{x}-, \hat{y}+, \hat{z}-)$
N_4	$(\hat{x}+, \hat{y}-, \hat{z}-)$	$(\hat{x}-, \hat{y}+, \hat{z}+)$
N_5	$(\hat{x}-, \hat{y}+, \hat{z}+)$	$(\hat{x}+, \hat{y}-, \hat{z}-)$
N_6	$(\hat{x}-, \hat{y}+, \hat{z}-)$	$(\hat{x}+, \hat{y}-, \hat{z}+)$
N_7	$(\hat{x}-, \hat{y}-, \hat{z}+)$	$(\hat{x}+, \hat{y}+, \hat{z}-)$
N_8	$(\hat{x}-, \hat{y}-, \hat{z}-)$	$(\hat{x}+, \hat{y}+, \hat{z}+)$

Similarly, the number of pairs of particles for which S. $\hat{x}$ is $(+)$ in beam 1 and S. $\hat{z}$ is $(-)$ in beam 2 are $(N_1 + N_3)$, and the number of pairs of particles for which S. $\hat{z}$ is $(-)$ in beam 1 and S. $\hat{y}$ is $(-)$ in beam 2 is $(N_2 + N_6)$. Since all the populations are positive numbers, the following inequality is obtained for the populations of the pairs of particles in the two beams:

$$N_1 + N_2 \leq (N_1 + N_3) + (N_2 + N_6) \tag{7.85}$$

The probability that on random selection a particle in beam 1 registers $(+)$ for S. $\hat{x}$ and a particle in beam 2 registers $(-)$ for S. $\hat{y}$ measurement is:

$$P(\hat{x}+, \hat{y}-) = \frac{N_1 + N_2}{\sum_{i=1}^{8} N_i} \tag{7.86}$$

Similarly, other probabilities can be described as:

$$P(\hat{x}+, \hat{z}-) = \frac{N_1 + N_3}{\sum_{i=1}^{8} N_i} \tag{7.87}$$

$$P(\hat{z}-, \hat{y}-) = \frac{N_2 + N_6}{\sum_{i=1}^{8} N_i} \tag{7.88}$$

Thus, the inequality in Equation (7.85) now becomes:

$$P(\hat{x}+, \hat{y}-) \leq P(\hat{x}+, \hat{z}-) + P(\hat{z}-, \hat{y}-) \tag{7.89}$$

The previous equation is called Bell's inequality. This inequality for the probabilities is obtained by assuming local correlations between the particles. If experimentally this inequality is violated, then the correlations between the particles are much more sophisticated than their classical counterparts. Several experiments were performed to test Bell's inequality. In 1982, the experiments by Alain Aspect and collaborators confirmed violation of the inequality. Their results were in excellent agreement with the predictions of quantum mechanics. Their experiments were further refined and improved, but the results remained the same.

8 Quantum Computer

8.1 CLASSICAL COMPUTER

At present computers are an intrinsic part of our day to day life. It is difficult to even imagine life without them. The "processing and storage of information" functions of the computer have greatly accelerated the pace of "human progress." Looking back one hundred and fifty years there were no cars, airplanes, computers, and smart phones etc., it was a totally different time. Because of the ubiquitousness and utility of these marvelous machines, access to the information from and influence on the entire world is now readily within our reach. The invention of the computer and the accompanying "digital revolution" has driven much of this progress.

What is a computer?

A computer is a machine that stores and processes information. For example, a certain amount of data is input into a computer, it processes the data by performing mathematical and logical operations and provides final output data. It simply uses the information contained in the input data and translates into more meaningful or useful output data. All mathematical calculations, graphing, and storing of information can be performed on a computer. In addition, one computer can be connected to the other computers using various telecommunication technologies. This permits the transfer of information between them. By connecting all (or many of) the computers in the world, information can be transferred from one side of the world to the other side of the world without even stepping outside of one's room.

What are the basic principles of computer function?

There are two principal components of a computer, a memory unit and a Central Processing Unit (CPU). A memory unit is a system that stores data and instructions. In memory, data is stored as "bits." A bit is the smallest unit of information and can be either a 1 or a 0. Bits are stored into sets of eight bits called "bytes," and bytes can be organized into sets of 4 or 8 bytes called "words." In the CPU, arithmetic and logical operations are performed on the data by a set of instructions called a program unit. This set of instructions is stored in the memory. A set of instructions for performing a task is called an algorithm. For example, to add two numbers, a basic summing algorithm is applied. In modern computers, circuits are used to perform a task applying an algorithm. A circuit involves input/output bits, wires and logical gates which carry information and perform computational tasks. A logical gate is a device that performs operations on one or more bits and produces one or more output bits. The NOT and AND logic gates are described below in Table 8.1. For example, the action of a NOT gate on a bit is to flip its state, such that if the input bit is 0 then a NOT gate transforms it into 1 and vice versa (Figure 8.1).

To solve any mathematical or computational problem using a computer, an "efficient algorithm" for solving the problem is required. In fact, any problem can be solved using a classical computer. How long does it take to solve a problem? The amount of time or number of steps an algorithm requires to solve a problem is a reasonable measure of the efficiency of the algorithm. If the time or the number of steps required by an algorithm to solve a problem is infinite or exponential, then the algorithm is reasonably considered to be "inefficient." If the time required follows a polynomial function of time, then the algorithm is considered to be "efficient." The magnitude of the processing resources, for example, the number of CPU operations, or "FLOPS," that is, Floating Point Operations Per Second, required is a direct measure of the "effort required" by the computer to solve a given problem. If the resources required is infinite or very large, then the whole process of solving the problem may have to be abandoned. Therefore, not only the time required but also the CPU operations required must be within reasonable limits to solve any computing problem. Computational

TABLE 8.1

Truth Tables for NOT, AND and OR Gates

NOT

Input	Output
0	1
1	0

AND

Input A	Input B	Output AB
0	0	0
0	1	0
1	0	0
1	1	1

OR

Input A	Input B	Output A+B
0	0	0
0	1	1
1	0	1
1	1	1

complexity is the study, or a measure of the time and processing resources required to solve complex problems. The main task of such a field is to prove the minimum bounds on the resources required by the best possible algorithm for solving a given computational problem. Thus, a problem is regarded as easy and feasible if the algorithm applied for solving the problem requires polynomial time (t^n) and resources. It is considered difficult and not practical if the time or processing resources required is infinite or exponential. According to this criterion, the computational problem of factoring a very large integer into its prime factors is not feasible or practical in a reasonable amount of time with a classical computer. The number of steps required grows exponentially with the size of the number. There are many more such problems which are intractable on a classical computer. This should encourage a search for new approaches of solving these problems computationally. A (successful) quantum computer would be a huge step towards such a goal.

In modern computers, all the above gate operations are performed using transistors. Use of transistors as hardware for computers began in the late 1950s and 1960s. Since then, the size of the transistor has been dramatically reduced and the overall complexity and processing power of integrated circuits (with many transistors) has been rapidly increasing. According to "Moore's law" (a very

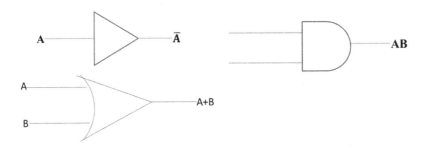

FIGURE 8.1   The schematic above illustrates symbols for NOT, AND and OR gates.

useful planning tool for the rapid advancement of the industry rather than an actual "law"), computing power doubles at constant cost roughly once every two years. Transistors will continue to shrink in size to a point that conventional approaches of the fabrication of computing devices will run up against difficulties of size and quantum limits. These quantum limits include the thickness of the dielectrics (insulators) and even conductors in transistors and other devices approaching the width of single atomic layers and their manifested electric field consequences of the layers and conductor lines being mismatched with their physical dimensions due to these quantum effects. These and other quantum effects have already significantly interfered in the functioning of electronic devices as they are being scaled to smaller dimensions and have become a major obstacle to the continuance of Moore's Law. One possible solution to these challenges is a computing device that functions on the principles of quantum mechanics, that is, "a quantum computer."

Conceptual Question 1: Explain, why we need a quantum computer. Describe in your own words the main advantages of a quantum computer?

8.2 QUANTUM COMPUTER

A quantum computer is a device that can perform computational tasks based on the principles of quantum mechanics.

What are the advantages of such a device over a classical computer? To answer such a question, it would be appropriate to review some of the fundamental principles of quantum mechanics (Figure 8.2).

Superposition principle

The quantum state of a particle can be described by a finite or infinite linear superposition of states. Consider a particle in an infinite square well. The state of such a particle can be described as a linear

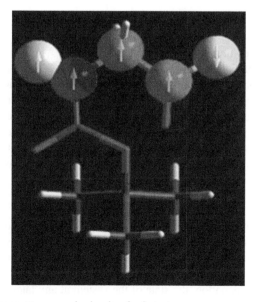

FIGURE 8.2 A quantum computer comprised only of a few atoms.

superposition of infinite states. Before any measurement is performed, the particle exists in such a state. Such a state can be described as,

$$\psi(x,t) = \sum_{n=1}^{\infty} c_n \varphi_n(x) e^{-iE_n t} \tag{8.1}$$

Let us say that each of the infinite states carries a bit of information, and a computational task is performed on such a state. A computational task on such a state of a particle, will perform operations on all these infinite states simultaneously. Which means that employing a "single particle" (the smallest resource that can currently be conceived of), a computational operation can be performed in parallel on an infinite number of states. Thus, the Superposition Principle permits computations as if one were to be able to access an infinite number of parallel processors with the minimum conceivable number of resources. In an actual quantum computer, parallel processing can indeed be performed using a minimum of resources which is impossible in a classical computer. To perform parallel processing in a classical computer, huge resources and space are required. However, when a measurement is performed on quantum bit (or "qubit"), the quantum superposition state collapses into a single state and only delivers the information equivalent of a "single bit" just like a classical computer. However, this superposition state can be used to extract the global properties of a system without any need for multiple qubits, which would be required of a classical computer. The other advantages of a quantum computer are reversible gate operations, and entanglement that delivers the effect of an exponential increase in the information processing power of a quantum computer as compared to its classical computer counterpart. These properties will be discussed in more detail later in the chapter.

8.2.1 QUBIT

Consider the case of a spin-1/2 particle in a linear superposition state according to the equation:

$$|\Psi\rangle = c_\uparrow |\uparrow\rangle_z + c_\downarrow |\downarrow\rangle_z \tag{8.2}$$

Consider that the spin-up state of the particle is represented as a classical bit "0," and the spin-down state as classical bit "1." The equation for the state of the particle can be written as:

$$|\Psi\rangle = c_0 |0\rangle + c_1 |1\rangle \tag{8.3}$$

where c_0 and c_1 are complex number coefficients.

The above state of the particle is called a qubit (quantum bit). A qubit can exist in a continuum of states as above until it is observed. A qubit is a smallest unit of information in a quantum computer similar to a bit in a classical computer. It can hold up to two bits of information using "superdense coding." This will be discussed in more detail later in the chapter. In the above state c_0 and c_1 are unknown coefficients. If a measurement is performed on the qubit to determine its state, it will collapse into either state $|0\rangle$ or $|1\rangle$. The absolute square of these coefficients yields the probability of observing either state $|0\rangle$ or $|1\rangle$ upon measurement, and these coefficients add up to unity:

$$|c_0|^2 + |c_1|^2 = 1 \tag{8.4}$$

There is an infinity of complex numbers whose absolute squares can add up to unity as in the equation above. How are these coefficients for a state determined theoretically? As was discussed in the Chapter 3, these coefficients can be determined by solving the Schrodinger equation for a system. All that is necessary to know is the initial state of the system. If initial state is known, then by applying the solution of the Schrodinger equation, the state of the system at any later time can be determined.

How many bits of information are carried by qubits?

For a spin-1/2 system, these coefficients are:

$$|c_0| = |c_1| = \frac{1}{\sqrt{2}} \tag{8.5}$$

Therefore, if there are N particles prepared in the state of (Equation 8.3), and measurements are performed to determine their state, $N/2$ of the particles will collapse to state $|0\rangle$ and the remaining $N/2$ will collapse to state $|1\rangle$. In other words, if 10 qubits are measured, then 5 of them will be found in the $|0\rangle$ state and the remaining 5 will be found in the $|1\rangle$ state. Thus, 10 qubits collapse to 10 classical bits upon measurement. But this does not mean that the amount of information carried by 10 qubits is same as that carried by 10 bits. The minimum amount of information carried by a qubit is two bits, and 10 qubits carry about 10×2 bits of information if they exist as independent qubits (no interaction between the qubits), and 2^{10} bits of information if they are all in a superposition state.

If the above coefficients are unknown and can have any arbitrary values, and yet their absolute squares satisfy Equation (8.4), then how are the values of such coefficients to be determined? There's no way of determining such coefficients in theory. So, we cannot tell anything about the amount of information carried by an assembly of qubits. The only way to determine how much information is carried by a number of qubits is to perform measurements and to determine the processing power of the qubits by the result of the measurements. One would need to prepare many samples of (let say) N qubits in such a state and then perform measurements. If each sample give the same results, then the state of the qubit and their processing power would be known. For example, consider 10 samples of N qubits prepared in an unknown state. If the outcome of a measurement on each sample shows that $N/2$ qubits collapses to state $|0\rangle$ and remaining qubits collapse to state $|1\rangle$, then, the unknown state has coefficients as defined by Equation (8.5). Therefore, to determine an unknown state experimentally, one would need many such samples. For an unknown state, one would need an infinite number of such samples of qubits to determine such a state experimentally.

This requirement for an "infinite" number of qubits renders it impossible to determine the unknown state of a qubit. So, if we do not know the state of the qubit, then how can computations be performed on it? Interestingly, there is a way around this problem. If we know the initial state of a qubit, then theoretically these coefficients can be determined at any later time. Therefore, for such a system, infinite number of qubits will not be required to determine the state. Such a situation is called a "known state." Thus, to perform computations on a qubit, it is necessary that a qubit be prepared initially in a known state.

All these points can be illustrated using a Bloch vector. A Bloch vector representation of a qubit is a geometric representation and is a very useful method of visualizing a qubit.

As discussed above, since the coefficients c_0 and c_1 are complex numbers, the state of a qubit can be written as:

$$|\psi\rangle = e^{i\gamma}\left(\cos\frac{\theta}{2}|0\rangle + \sin\frac{\theta}{2}e^{i\varphi}|1\rangle\right) \tag{8.6}$$

where θ, γ and φ are real numbers. The factor $e^{i\gamma}$ at the beginning of the equation can be removed since it has no observable effects, and for that reason the following equation is obtained:

$$|\psi\rangle = \cos\frac{\theta}{2}|0\rangle + \sin\frac{\theta}{2}e^{i\varphi}|1\rangle \tag{8.7}$$

The numbers θ and φ define a point on the unit three-dimensional sphere, as described in Figure 8.3. This sphere is called the Bloch sphere. There are an infinite number of points on the Bloch sphere, and any state of the qubit can be represented on it. A qubit in an unknown state can be represented on any point of the Bloch sphere. However, if the initial state of the qubit is known, then the manner in which the state of qubit changes with time can be observed by the path traced by a qubit on

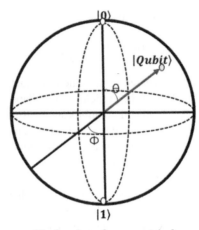

**Bloch sphere for geometrical
representation of a qubit**

FIGURE 8.3 A Bloch sphere representation of a qubit.

the Bloch sphere. It is a very useful visualization tool that can be used to understand the concepts of quantum information and quantum computation. However, for multiple qubits it becomes increasingly difficult to use such a tool as the number of qubits is increased.

Conceptual Question 2: How many minimum bits of information can be stored in a single qubit?

Conceptual Question 3: Why do we need to prepare a qubit in a known state to perform computations on it?

8.2.2 MULTIPLE QUBITS

The state of two qubits can be described as a tensor product:

$$|\psi\rangle = (\alpha|0\rangle_1 + \beta|1\rangle_1) \otimes (\gamma|0\rangle_2 + \mu|1\rangle_2)$$
$$= c_{00}|00\rangle_{12} + c_{01}|01\rangle_{12} + c_{10}|10\rangle_{12} + c_{11}|11\rangle_{12} \tag{8.8}$$

In simple notation, the two qubits state can be written as:

$$|\psi\rangle = c_{00}|00\rangle + c_{01}|01\rangle + c_{10}|10\rangle + c_{11}|11\rangle \tag{8.9}$$

Two qubits exist as a superposition of four computational basis states ($|00\rangle$, $|01\rangle$, $|10\rangle$, $|11\rangle$). Whereas, in a classical computer, eight qubits will be required to store four computational basis states. The normalization condition for the above qubit requires that:

$$\sum_{i=0,j=0}^{1,1} |c_{ij}|^2 = 1 \tag{8.10}$$

Similarly, to the case of a single qubit, measurements on two qubits will result in their collapse to a single two qubit state as one of the above computational basis states. The post measurement state will be one of those states. However, measurement on one of the qubits (e.g., the first qubit as "0"), the post measurement state will result in state:

$$|\psi'\rangle = \frac{c_{00}|00\rangle + c_{01}|01\rangle}{\sqrt{|c_{00}|^2 + |c_{01}|^2}} \qquad (8.11)$$

The post measurement state is renormalized with a factor in the denominator, so that it satisfies the normalization condition. In the section on measurement (below), it will be shown how measurement results in the state shown above in Equation (8.11).

8.2.2.1 Entanglement

One of the most important two qubit states is the Bell state or the EPR pair. Such a state is defined as:

$$|\psi\rangle = \frac{|00\rangle + |11\rangle}{\sqrt{2}} \qquad (8.12)$$

The above state is not a tensor product state as shown in Equation (8.8). The EPR (Einstein-Podolsky-Rosen) paradox was discussed in Chapter 7. Their experiment demonstrated that a quantum mechanical wavefunction does not provide a complete description of physical reality. Later in 1964, J.S Bell, proved that the EPR argument was incompatible with quantum mechanics.

The above state is very interesting for quantum computation and is mainly responsible for quantum teleportation and superdense coding. In the above state, by performing a measurement on one of the qubits, the state of the other qubit can be determined. If a measurement on qubit 1 is performed and it is found to be in state $|0\rangle$, then qubit 2 will also be in state $|0\rangle$. The post measurement state will be $|00\rangle$. As a result, measurement on the second qubit yields the same result as the first. Thus, by determining the state of the first qubit, the state of second qubit can be determined without performing any measurements on the second. That is, the measurement outcomes are correlated. These correlations are very strong, and such strong correlation could never exist in classical systems. These correlations are also called entangled states. Such correlations between the qubits exists in quantum computers and are responsible for making quantum computers more powerful than classical computers.

Consider an "n" qubit multiple state, $|x_1, x_2, x_3, x_4 \ldots\rangle$. Such a multiple qubit state can be expressed with a 2^n computational basis and has 2^n amplitudes. This means that 2^n complex numbers can be stored by using only "n" qubits. For example, consider $n = 1000$, where the number of complex numbers are 2^{1000}, which is a remarkably a large number. In a classical computer, n bits can store only n computational basis states at a time. Storing this many complex numbers on a classical computer is simply impossible. This phenomenon is called quantum parallelism, and this may result in a quantum computer delivering an exponential increase in processing power as compared to a classical computer.

Conceptual Question 4: What is quantum parallelism? Explain.

8.2.3 QUBIT GATES

Classical computers consist of wires and logic gates. The wires are used to carry information around the circuit, while the logic gates perform computational operations for the manipulations of information, converting it into output. Similarly, a quantum computer is built from a quantum circuit that consists of quantum gates and wires. Quantum gates perform the computational operations on the information. Though the functions of classical and quantum gates are the same, they are fundamentally distinct in their output characteristics. All quantum gates perform unitary operations and are reversible. Unlike classical gates, which are not reversible except for Toffolli gates. Reversibility of quantum gates offer a great advantage, which allows, any gate operation to be reversed without any loss of information.

Quantum gates are represented by unitary matrices and an "n" number of qubits gate can be described by $2^n \times 2^n$ matrix. Due to the unitary nature of the logic gate, it is a reversible operation. The quantum gate acts linearly, which means that it converts one state to another state in which the role of the states is interchanged, but the amplitude of the state remain the same.

i. For the example of a NOT gate,
The gate operation can be described as,

$$\text{Input state} \rightarrow |\Psi\rangle = c_0|0\rangle + c_1|1\rangle \rightarrow (\text{gate operation})|\Psi\rangle = c_0|1\rangle + c_1|0\rangle$$
$$= \text{Output state} \tag{8.13}$$

The input state is changed to a reversed input state by a linear NOT gate operation. In matrix form, the NOT gate can be described as:

$$X = \begin{pmatrix} 0 & 1 \\ 1 & 0 \end{pmatrix} \tag{8.14}$$

$$X|\Psi\rangle = \begin{pmatrix} 0 & 1 \\ 1 & 0 \end{pmatrix}\begin{pmatrix} c_0 \\ c_1 \end{pmatrix} = \begin{pmatrix} c_1 \\ c_0 \end{pmatrix} \tag{8.15}$$

All matrices for the gate operation must be unitary such that:

$$XX^\dagger = 1 \tag{8.16}$$

This requirement is the only constraint of quantum gate operation. Thus, all quantum gates can be represented by unitary matrices.

ii. Z (Phase gate):
This gate allows the $|0\rangle$ state to remain unchanged and flips the sign of the $|1\rangle$ state to $-|1\rangle$:

$$Z = \begin{pmatrix} 1 & 0 \\ 0 & -1 \end{pmatrix} \tag{8.17}$$

$$Z|\Psi\rangle = \begin{pmatrix} 1 & 0 \\ 0 & -1 \end{pmatrix}\begin{pmatrix} c_0 \\ c_1 \end{pmatrix} = \begin{pmatrix} c_0 \\ -c_1 \end{pmatrix}$$

$$ZZ^\dagger = 1 \tag{8.18}$$

iii. Hadamard gates:
This gate turns the state $|0\rangle$ to a linear superposition state $((|0\rangle + |1\rangle)/\sqrt{2})$, and state $|1\rangle$ to $(|0\rangle - |1\rangle)/\sqrt{2}$. The unitary matrix for such a gate operation is:

$$H = \frac{1}{\sqrt{2}}\begin{pmatrix} 1 & 1 \\ 1 & -1 \end{pmatrix} \tag{8.19}$$

This gate is one of the most useful gates, since it transforms a classical bit into a qubit. The twice operation on this bit leads to the same initial state of the bit, $H^2 = I$. By applying the Hadamard operation twice on a $|0\rangle$ bit yields a $|0\rangle$ bit.

iv. C-NOT:

The C-NOT gate or the "controlled NOT" gate is a two-qubit gate. It has two input qubits, one of which is called a control qubit and the other is called a target qubit. If the control bit is set to $|0\rangle$, then the target qubit is unchanged. If the control qubit is $|1\rangle$ then the target qubit is flipped. This gate operation results in an entangled state. The two qubit C-NOT gate is described below.

In the Figure 8.4, the qubit on the top is the control bit, and the one at the bottom is the target qubit. So if the control qubit is a superposition state according to: $|\psi\rangle = (1/\sqrt{2})|0\rangle_1 + (1/\sqrt{2})|1\rangle_1$ and the target bit is $|0\rangle_2$, the C-NOT gate operation will result in a Bell state or EPR pair: $(|00\rangle + |11\rangle)/\sqrt{2}$. All the known Bell states can be produced using C-NOT gate. These states are:

$$|\psi_{00}\rangle = \frac{|00\rangle + |11\rangle}{\sqrt{2}} \tag{8.20}$$

$$|\psi_{01}\rangle = \frac{|01\rangle + |10\rangle}{\sqrt{2}} \tag{8.21}$$

$$|\psi_{10}\rangle = \frac{|00\rangle - |11\rangle}{\sqrt{2}} \tag{8.22}$$

$$|\psi_{11}\rangle = \frac{|01\rangle - |10\rangle}{\sqrt{2}} \tag{8.23}$$

The C-NOT gate and single qubit gates can be used to generate any multiple qubit gates. The proof of this statement will not be presented here. Rather this statement will be accepted and applied for the development of any type of gates. The truth tables for all the above gates are shown in Table 8.2.

Conceptual Question 5: Explain the differences between a classical gate and quantum gate?

PROBLEMS

8.1 Show that the state of two qubit, $|01 \neq |10\rangle$, and is non-commutative.

8.2 Show that the Hadamard gate matrix (Equation 8.17) is a unitary matrix.

8.3 Explain the difference between the tensor product state of two qubits and the entangled state (Equation 8.12).

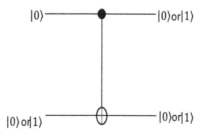

FIGURE 8.4 This figure illustrates a two qubit C-NOT gate. The second qubit is control qubit.

TABLE 8.2

Truth Tables of Quantum NOT and C-NOT Gates

NOT

Input	Output
$|0\rangle$	$|1\rangle$
$|1\rangle$	$|0\rangle$

C-NOT

Control	Target	Output
$|\Psi\rangle = c_0|0\rangle + c_1|1\rangle$	$|0\rangle$	$|\Psi\rangle = c_0|00\rangle + c_1|11\rangle$
$|\Psi\rangle = c_0|0\rangle + c_1|1\rangle$	$|1\rangle$	$|\Psi\rangle = c_0|01\rangle + c_1|10\rangle$

8.3   QUANTUM ALGORITHMS

A quantum algorithm is a step by step set of instructions that applies special quantum features such as the superposition principle and entanglement to perform computational operations using realistic quantum models such as quantum circuits. A quantum algorithm can only be run on a quantum computer. However, all classical algorithms can be run on a quantum computer. The very first quantum algorithm was suggested by David Deutsch. In his paper in 1981, he suggested a quantum algorithm based on quantum parallelism, and using such an algorithm a global property of a mathematical function can be determined using a single quantum circuit. For example, to evaluate a function $f(x)$, for $x = 0$ or $x = 1$, at least two different circuits are required on a classical computer. But using a single quantum circuit, a global property of $f(x)$, as $f(0) \oplus f(1)$, (where $\oplus$ represent addition modulo 2) can be determined.

i. Deutsch's algorithm

Consider the circuit illustrated in Figure 8.5.

In this circuit, a Hadamard operation is applied to state $|0\rangle$, to transform it into the state $(|0\rangle + |1\rangle)/\sqrt{2}$, then a U_f operation ($U_f|x\rangle|y\rangle = |x\rangle|y \oplus f(x)\rangle$) is applied resulting in state:

$$\frac{|0, f(0)\rangle + |1, f(1)\rangle}{\sqrt{2}} \tag{8.24}$$

This state is interesting, since it seems as if two values of the function $f(x)$ are evaluated simultaneously. If $f(0) = f(1)$, the first qubit is in state $(|0\rangle + |1\rangle)/\sqrt{2}$, and then by applying the Hadamard operation on this qubit again, the output will be $|0\rangle$ with certainty. However, if $f(0) \neq f(1)$, then if the first qubit is measured to obtain $|1\rangle$ then that function is known to be "balanced" (i.e., that function has a value of 0 for half of the values of x), and the output will be declared to be "balanced." Thus, we have an algorithm that can

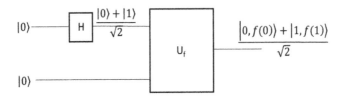

FIGURE 8.5   The schematic illustrates a quantum circuit for operation of Deutsch's algorithm.

determine or "guess" the probability of the global property of a function being true. With a classical computer such a "guessing" would not be possible. This can be illustrated further with the following equations. Consider two qubits in states $|0\rangle$ and $|1\rangle$. By applying a Hadamard operation to the first and second qubits the following state is obtained:

$$|\psi_1\rangle = \left[\frac{|0\rangle + |1\rangle}{\sqrt{2}}\right]\left[\frac{|0\rangle - |1\rangle}{\sqrt{2}}\right] \qquad (8.25)$$

By applying a U_f operation to the above state: $(U_f|x\rangle((|0\rangle - |1\rangle)/\sqrt{2}) \rightarrow (-1)^{f(x)}|x\rangle((|0\rangle - |1\rangle))/\sqrt{2})$ the following relation is obtained:

$$|\psi_2\rangle = \left[\frac{(-1)^{f(0)}|0\rangle}{\sqrt{2}}\left[\frac{|0\rangle - |1\rangle}{\sqrt{2}}\right] + \frac{(-1)^{f(1)}|1\rangle}{\sqrt{2}}\left[\frac{|0\rangle - |1\rangle}{\sqrt{2}}\right]\right] \qquad (8.26)$$

Here two possibilities are presented, that is, either $f(0) = f(1)$ or $f(0) \neq f(1)$ and for these two possibilities the following is derived:

$$|\psi_2\rangle = \pm\left[\frac{|0\rangle + |1\rangle}{\sqrt{2}}\right]\left[\frac{|0\rangle - |1\rangle}{\sqrt{2}}\right] \quad \text{for} \quad f(0) = f(1)$$

$$\qquad (8.27)$$

$$|\psi_2\rangle = \pm\left[\frac{|0\rangle - |1\rangle}{\sqrt{2}}\right]\left[\frac{|0\rangle - |1\rangle}{\sqrt{2}}\right] \quad \text{for} \quad f(0) \neq f(1)$$

By applying a Hadamard operation again to the first qubit for both possibilities, the following can be derived:

$$|\psi_3\rangle = \pm|0\rangle\left[\frac{|0\rangle - |1\rangle}{\sqrt{2}}\right] \quad \text{for} \quad f(0) = f(1)$$

$$\qquad (8.28)$$

$$|\psi_3\rangle = \pm|1\rangle\left[\frac{|0\rangle - |1\rangle}{\sqrt{2}}\right] \quad \text{for} \quad f(0) \neq f(1)$$

Thus, if the measurement outcome on the first qubit yields a $|0\rangle$, then with certainty it is known that the function is a constant. If the measurement outcome is $|1\rangle$, then with certainty it is known that the function is balanced. It is very interesting (and useful) to observe that a single quantum circuit can determine the global property of a given function.

ii. Deutsch's-Jozsa algorithm

This algorithm is a generalization of the Deutsch's algorithm above. The problem is to use an algorithm to determine whether a function $f(x)$ is constant for all values of x, or is balanced, such that it is equal to 1 for exactly half the of values of x, and 0 for the remaining half. Consider the problem of Alice and Bob who are located in different places in the world and trying to communicate with each other. Alice selects a number "x" from a list of 0 to $2^n - 1$ numbers and mails it to Bob. Bob has two functions, a constant function ($f(x)$) whose values is constant for all values of "x," and another function $g(x)$, that yields a value of 0 for half of the values of "x" and a value of 1 for the remaining half of the values of "x." He chooses one of these functions and calculates the value of the function for the number that Alice has sent to him. He then sends back the value of the function to Alice. Now, Alice needs to find out whether Bob has chosen $f(x)$ or $g(x)$. How many times does she need to query (i.e., to keep sending numbers) Bob to find out which function Bob is applying to her number? Classically, with the best deterministic algorithm she needs to query Bob $2^{n-1} + 1$ times. Because she may receive 2^{n-1} 0's before receiving a 1 to determine that the function is $g(x)$. For this, Alice will be sending "n" bits to Bob to determine the function. Interestingly, by using qubits, instead of classical bits, and if Bob calculates the function using the Unitary transform (U_f) as discussed in the previous algorithm, then Alice could determine the function in a single correspondence with Bob. Further details of this algorithm will not be discussed since it is beyond the scope of this chapter. However, from this simple example, some of the promise of quantum computing for the solution of large intractable problem can be readily discerned in comparison to its classical computing counterpart. The past several decades of research in this field has resulted in four classes of quantum algorithms that have been known to solve problems more efficiently than any known classical algorithm. These are, (i) Quantum algorithms based on the Fourier transform such as the Deutsch-Jozsa algorithm, (ii) Shor's algorithm for factoring a prime number, and (iii) Quantum search algorithms, and (iv) Quantum simulation, to simulate a quantum mechanical system. The main questions whose answers will continue to evolve are: What are the problems for which only quantum algorithms will offer efficient solutions? What are the overall distinguishing features of quantum algorithms as compared to their classical counter-parts?

8.4   QUANTUM MEASUREMENT

In the previous chapters, it was demonstrated that the Schrodinger equation is a deterministic equation. The wavefunction solution of the Schrodinger equation is also a deterministic function. This means that just like any classical function, if the initial wavefunction of a system is known, then the wavefunction for any future time can be determined. The wavefunction evolves unitarily with time. However when measurements are performed to determine what is occurring inside the system, the wavefunction collapses into one of the possible states and stops evolving unitarily. Theoretically, only probabilistic information about the system can be extracted from the wavefunction. To understand the effect of measurement on a quantum system, consider a single qubit:

  i. Single qubit

$$|\Psi\rangle = c_0|0\rangle + c_1|1\rangle$$

Let quantum measurements be described by measurement operators such as $\{M_m\}$, and index m (0, 1) which refers to the measurement outcome, such that the measurement operator M_0 causes collapse to state $|0\rangle$ and so on. The probability that a result m occurs is described by:

$$p(m) = \langle\Psi|M_m^\dagger M_m|\Psi\rangle \tag{8.29}$$

where $M_0 = |0\rangle\langle0|$, and $M_1 = |1\rangle\langle1|$ and are Hermitian operators.

The state of the system after the measurement is described as:

$$|\Psi'\rangle = \frac{M_m|\Psi\rangle}{\sqrt{\langle\Psi|M_m^\dagger M_m|\Psi\rangle}} \tag{8.30}$$

The measurement operators satisfy the completeness relation:

$$\sum_m M_m^\dagger M_m = M_0^\dagger M_0 + M_1^\dagger M_1 = 1 \tag{8.31}$$

The completeness relation also requires that all probabilities must add up to unity:

$$\sum_m p(m) = 1 \tag{8.32}$$

The probability that measurement collapses the state of the qubit to a state $|0\rangle$ is:

$$p(0) = \langle \Psi | M_0^\dagger M_0 | \Psi \rangle = |c_0|^2 \tag{8.33}$$

The state after the collapse is:

$$|\Psi'\rangle = \frac{M_0 |\Psi\rangle}{\sqrt{\langle \Psi | M_0^\dagger M_0 | \Psi \rangle}} = \frac{c_0}{|c_0|} |0\rangle \tag{8.34}$$

Similarly, the probability for collapse to state $|1\rangle$ is:

$$p(1) = \langle \Psi | M_1^\dagger M_1 | \Psi \rangle = |c_1|^2 \tag{8.35}$$

And the state after the collapse is:

$$|\Psi'\rangle = \frac{M_1 |\Psi\rangle}{\sqrt{\langle \Psi | M_1^\dagger M_1 | \Psi \rangle}} = \frac{c_1}{|c_1|} |1\rangle \tag{8.36}$$

The above measurement procedure can be applied to determine the probabilistic outcome and the state after the collapse, for the quantum states $|0\rangle$ and $|1\rangle$ that are orthogonal to each. Non-orthogonal states cannot be distinguished reliably, and thus the procedure above fails to provide the correct measurement probabilities if the quantum states are non-orthogonal. The above procedure can be extended to multiple qubits.

ii. Multiple qubits

Consider a two-qubit state:

$$|\Psi\rangle = c_{00}|00\rangle + c_{01}|01\rangle + c_{10}|10\rangle + c_{00}|11\rangle \tag{8.37}$$

For the above state, the measurement operators are:

$$M_{00} = |00\rangle\langle00|, \ M_{01} = |01\rangle\langle01|, \ M_{10} = |10\rangle\langle10|, \ M_{11} = |11\rangle\langle11| \tag{8.38}$$

Such that the identity relation is followed:

$$\sum_m M_m^\dagger M_m = M_{00}^\dagger M_{00} + M_{01}^\dagger M_{01} + M_{10}^\dagger M_{10} + M_{11}^\dagger M_{11} = 1 \tag{8.39}$$

And the probabilities of the states are:

$$\sum_{mm'} p(mm') = p(00) + p(01) + p(10) + p(11) = 1 \tag{8.40}$$

The probability that measurement collapses the state of the qubit to a state $|00\rangle$ is:

$$p(00) = \langle \Psi | M_{00}^\dagger M_{00} | \Psi \rangle = |c_{00}|^2 \tag{8.41}$$

The state after the collapse is:

$$|\Psi'\rangle = \frac{M_{00}|\Psi\rangle}{\sqrt{\langle\Psi|M_{00}^\dagger M_{00}|\Psi\rangle}} = \frac{c_{00}}{|c_{00}|}|00\rangle \tag{8.42}$$

The above method requires measurement on both qubits simultaneously. In a real scenario multiple qubit are comprised of many qubits and it is almost impossible to perform simultaneous measurements on all qubits. Therefore, it is necessary that measurement be performed on only one of the qubits. The measurement operators for such a process are:

$$M_0^1 = |00\rangle\langle00| + |01\rangle\langle01|, \quad M_1^1 = |10\rangle\langle10| + |11\rangle\langle11| \tag{8.43}$$

Here the measurement operator M_0^1 represent measurement on the first qubit when it is in state $|0\rangle$, and M_1^1 represent measurement on a first qubit when it in state $|1\rangle$. The probability of measuring the first qubit in state $|0\rangle$ is:

$$p_0^1 = \langle\Psi|M_0^{1\dagger}M_0^1|\Psi\rangle = |c_{00}|^2 + |c_{01}|^2 \tag{8.44}$$

The state after the measurement is:

$$|\Psi'\rangle = \frac{M_0^1|\Psi\rangle}{\sqrt{\langle\Psi|M_0^{1\dagger}M_0^1|\Psi\rangle}} = \frac{c_{00}|00\rangle + c_{01}|01\rangle}{\sqrt{|c_{00}|^2 + |c_{01}|^2}} \tag{8.45}$$

Such a measurement causes collapse to the above state and excludes the other states $|10\rangle$ and $|11\rangle$.

8.5   DENSITY OPERATOR

In the previous chapters, a wavefunction was applied to describe a quantum state. Another approach for describing the quantum state is known as a density operator or density matrix method of a quantum state. This approach is very useful for describing a large ensemble of quantum states. Consider a quantum system that is comprised of states $|\phi_1\rangle, |\phi_2\rangle \ldots, |\phi_n\rangle$ with respective probabilities $p_1, p_2, \ldots p_n$. The density operator for such a system is:

$$\rho = \sum_i p_i|\phi_i\rangle\langle\phi_i| \tag{8.46}$$

In the above equation, the state of a quantum system is described in terms of the probabilities for a state. The above state is called a "mixed state," since the system is in a mixture of different states in the ensemble of ρ. Such a state cannot be described directly by using the wavefunction approach. Such a state arises when interference terms of a pure state vanish.

A qubit state, $|\Psi\rangle = c_0|0\rangle + c_1|1\rangle$, can be described with a density operator as:

$$\rho = |\psi\rangle\langle\psi| = |c_0|^2|0\rangle\langle0| + c_0c_1*|0\rangle\langle1| + c_1c_0*|1\rangle\langle0| + |c_0|^2|1\rangle\langle1| \tag{8.47}$$

Such a state is called a "pure state." Such a state is equivalent to a superposition state of a wavefunction. In the generalized form, the pure state can be written as:

$$\rho = |\psi\rangle\langle\psi| = \sum_{j,k} c_jc_k*|j\rangle\langle k| \tag{8.48}$$

If Equation (8.47) applies, when the second and third term vanishes, the pure state turns into a mixed state.

EXAMPLE 8.1

Consider a qubit state, $|\Psi\rangle = (1/\sqrt{2})|0\rangle + (1/\sqrt{2})|1\rangle$, with a density operator,

$$\rho = |\psi\rangle\langle\psi| = \frac{1}{2}|0\rangle\langle 0| + \frac{1}{2}|0\rangle\langle 1| + \frac{1}{2}|1\rangle\langle 0| + \frac{1}{2}|1\rangle\langle 1| \tag{8.49}$$

The above is a pure state, and becomes a mixed state by the loss of interference terms:

$$\rho = \frac{1}{2}|0\rangle\langle 0| + \frac{1}{2}|1\rangle\langle 1| \tag{8.50}$$

It can be stated that the loss of phase information (loss of interference terms) transitions a pure state into a mixed state.

The time evolution of a closed quantum system can be described simply using a Unitary operator as:

$$\rho(t) = \sum_i p_i U(t)|\phi_i\rangle\langle\phi_i|U(t)^\dagger = U\rho U^\dagger \tag{8.51}$$

Measurements can also be described easily applying this approach. Consider a quantum system that is initially in a state $|\phi_i\rangle$, and a measurement is performed on such a system. The measurement probabilities for the outcome "m" can be described as:

$$p(m) = tr\left(M_m^\dagger M_m \rho\right) = \sum_i p_i tr(M_m^\dagger M_m |\phi_i\rangle\langle\phi_i|) \tag{8.52}$$

The density operator after the measurement of the state "m" can be described as:

$$\rho_m = \frac{M_m^\dagger \rho M_m}{tr\left(M_m^\dagger M_m \rho\right)} \tag{8.53}$$

8.5.1 Properties of a Density Operator

i. The trace of a density operator is always unity:

$$tr(\rho) = \sum_i p_i tr(|\phi_i\rangle\langle\phi_i|) = \sum_i p_i = 1 \tag{8.54}$$

ii. The density operator is a positive operator. For any arbitrary state $|\phi\rangle$, the overlap of this state on the density operator is:

$$\langle\phi|\rho|\phi\rangle = \sum_i p_i\langle\phi|\phi_i\rangle\langle\phi_i|\phi\rangle \geq 0 \tag{8.55}$$

iii. For a pure state,

$$tr(\rho^2) = 1 \tag{8.56}$$

iv. For a mixed state,

$$tr(\rho^2) < 1 \qquad (8.57)$$

8.5.2 REDUCED DENSITY OPERATOR

Consider a quantum system that is comprised of two sub-subsystems A and B. To gain information about one of the subsystems, a reduced density operator is commonly applied. It is an important tool for the analysis of composite quantum systems. The reduced density operator for a system A is defined by:

$$\rho^A = tr_B(\rho^{AB}) \qquad (8.58)$$

EXAMPLE 8.2

Consider a two-qubit state:

$$|\Psi\rangle = \frac{1}{\sqrt{2}}|00\rangle + \frac{1}{\sqrt{2}}|11\rangle \qquad (8.59)$$

And the density operator for such a state is:

$$\rho = |\psi\rangle\langle\psi| = \left(\frac{1}{\sqrt{2}}|00\rangle + \frac{1}{\sqrt{2}}|11\rangle\right)\left(\frac{1}{\sqrt{2}}\langle00| + \langle11|\right) \qquad (8.60)$$

The density operator of first qubit can be obtained by taking trace over the second qubit,

$$\rho^1 = tr_2(\rho) \qquad (8.61)$$

$$= \langle0|\rho|0\rangle + \langle1|\rho|1\rangle \qquad (8.62)$$

$$= \frac{1}{2}|0\rangle\langle0| + \frac{1}{2}|1\rangle\langle1| \qquad (8.63)$$

The density operator of the first qubit is in a mixed state $(tr(\rho^1\rho^1) < 1)$ while the density operator of the two qubits is in a pure state. This implies that the state of the two qubits is completely known in such a composite system. However, as a subsystem, the state of the qubit is not completely known. This is an interesting property of entangled quantum systems.

PROBLEMS

8.4   Show that the state $|\Psi\rangle = (1/\sqrt{2})|00\rangle + (1/\sqrt{2})|11\rangle$ is a pure state.

8.5   Consider the following density operator, $\rho = (1/3)|0\rangle\langle0| + (2/3)|1\rangle\langle1|$. Show that the same density operator can be described by the following basis vectors: $|a\rangle = \sqrt{1/3}|0\rangle + \sqrt{2/3}|1\rangle$, and $|b\rangle = \sqrt{1/3}|0\rangle - \sqrt{2/3}|1\rangle$.

8.6   DECOHERENCE

The wavefunction of a quantum system has complete information about the system and is described as a linear superposition of many quantum states. However, when a measurement is performed to determine the state of the system, it collapses the superposition state into one of the possible states. Thus, complete information about the system cannot be obtained. This behavior of quantum mechanics has puzzled scientists for several decades and is commonly referred to as a "measurement problem." Several scientists have attempted to explain such behavior theoretically. The very first and widely accepted explanation was provided by the "Copenhagen interpretation" (as discussed in Chapters 3 and 7). This interpretation was formulated by Neils Bohr (1928) and his co-workers. According to this interpretation, quantum systems do not actually exist in a definite state before the measurement, and the act of measurement causes the superposition state to collapse to a single state. In simple words, "quantum systems behave in this way because that is the way they are." It cannot explain what occurs in the measurement process that causes such a collapse of the wavefunction. There is another explanation known as "Many worlds Interpretation" which tries to describe the whole universe using a wavefunction. It was formulated by Hugh Everett III (1957). At present, it is not a widely accepted theory. Many scientists believe that this theory is inconsistent. The third theory that tries to explain the measurement process, is called the "Decoherence theory" and it is a relatively a new one. It is most useful in the field of quantum computation and quantum information. The foundation of this theory was laid by H. D Zeh, and W. D Zurek (1980). According to this theory, a quantum system interacts with its environment. The interaction between the quantum system and the environment causes it to lose information to the environment. This loss of information by the quantum system is referred to as decoherence. This also means loss of quantum coherence. It is an irreversible process, and once the information is lost, it cannot be recovered. The decoherence theory provides an explanation of the collapse of the wavefunction. For the last two decades this theory has become an important tool for the development of a reliable quantum computer.

Consider a qubit whose state is described by the following density operator:

$$\rho_{pure} = |c_0|^2|0\rangle\langle0| + c_0c_1{*}|0\rangle\langle1| + c_1c_0{*}|1\rangle\langle0| + |c_0|^2|1\rangle\langle1| \tag{8.64}$$

Such a state as discussed above is a pure state, and a linear superposition state. According to this theory, the interaction between this state and the environment causes it to lose information to the environment, and it degrades into a mixed state:

$$\rho_{mixed} = |c_0|^2|0\rangle\langle0| + |c_1|^2|1\rangle\langle1| \tag{8.65}$$

The time it takes for the system to lose information is referred to as the decoherence time (T_c). The states $|0\rangle$ and $|1\rangle$ are eigenstates with different energies. However, the process involved in the decoherence of a qubit does not involve energy exchange with the environment. This is mainly a loss of phase information and is referred to simply as a loss of coherence. The measurement outcome of the state will either lead to state $|0\rangle$ or $|1\rangle$ with a probability $|c_0|^2$ and $|c_1|^2$ respectively (Figure 8.6).

In a quantum computer, it is necessary that the time for performing any operation on a qubit (pure state) must be much greater than their decoherence time. Such decoherence times have been studied for many systems such as superconducting Josephson junctions, ion traps and atom interferometry. An understanding of the quantum-environment interaction or coupling can lead to the design of quantum computing systems whose decoherence times can be controlled and enhanced. This control over quantum systems to prevent their loss of information to the environment will likely be the key for the development of a reliable quantum computer. The next section of this chapter discusses the methods for overcoming decoherence in a quantum computer.

Conceptual Question 6: Explain the differences between an entangled state and a mixed state?

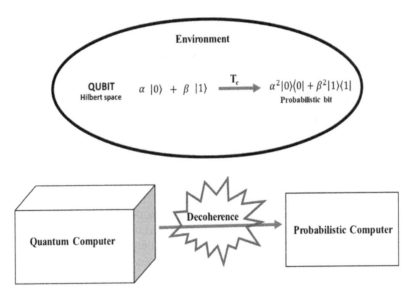

FIGURE 8.6 An illustration of decoherence of a qubit in a quantum computer. Such decoherence of qubits in a quantum computer causes it to turn into a probabilistic computer.

8.7 METHODS FOR OVERCOMING DECOHERENCE

It was discussed above that decoherence is the main obstacle in the way of building a reliable quantum computer. Many methods have been suggested to overcome decoherence. Some of the principle methods are:

- Quantum error-correction
- Decoherence-free subsystem/subspace
- Adiabatic computation

The first method is based on the classical approach of preventing errors in a classical computer. Perhaps counter-intuitively, such an approach is also possible for a quantum computer. This approach is discussed below.

8.7.1 QUANTUM ERROR-CORRECTING CODES

Classically, information carried by a bit is protected against errors by encoding it. At the end of the computation, the code is decoded, and the information carried by the bit is recovered. The method of quantum error-correcting codes is based on similar principles in the sense that it protects a qubit from errors as it is encoded and then followed by error correction and decoding. The schematic below describes such a process. However, the error correction is accomplished in such a way that the encoded state of the qubit is not disturbed by the measurement. A variety of quantum error-correcting codes exists, including early ones by P. Shor and A. M Steane. These require an error threshold on the order of 10^{-6}, which means they can tolerate one error per one million (10^{6}) gate operations. All these models are based on the assumption that errors are uncorrelated.

A quantum error correcting code is defined to be a unitary mapping of k qubits into a subspace of the quantum state space of n qubits such that if any arbitrary group of "t" qubits decoheres, the resulting $n - t$ qubits can be used to reconstruct the original quantum state of k encoded qubits. The decoherence time of the quantum state of k qubits is in general $1/k$ times that of a single qubit. Encoding a qubit into a code reduces overall decoherence time, however, by adding more qubits to the state, it is possible to detect and correct errors easily. This allows quantum error-correcting codes to be a useful approach for storing the information in the form of a code for an extended period of time.

Consider a qubit, $|\Psi\rangle = c_0|0\rangle + c_1|1\rangle$ that is encoded as a three-qubit quantum error correcting code:

$$|\Psi\rangle = c_0|000\rangle + c_1|111\rangle = c_0|0\rangle_L + c_1|1\rangle_L \tag{8.66}$$

where $|0\rangle_L = |000\rangle$, $|1\rangle_L = |111\rangle$ are logical $|0\rangle$ and $|1\rangle$, and not the physical zero and one states. The circuit for encoding such a state is described in Figure 8.7. The possible errors due to a single bit flip of any of the above three qubits will cause the following errors:

$$|\Psi\rangle_{\text{error}} = c_0|100\rangle + c_1|011\rangle \tag{8.67}$$

And a bit flip error caused by flipping of the first qubit:

$$|\Psi\rangle_{\text{error}} = c_0|010\rangle + c_1|101\rangle \tag{8.68}$$

Or a bit flip error caused by flipping of the second qubit:

$$|\Psi\rangle_{\text{error}} = c_0|001\rangle + c_1|110\rangle \tag{8.69}$$

Or a bit flip error caused by flipping of the third qubit.

All of these errors can be corrected by performing projection measurements on the three-qubit code. The projection measurement operators are:

$$P_0 = |000\rangle\langle000| + |111\rangle\langle111| \tag{8.70}$$

$$P_1 = |100\rangle\langle100| + |011\rangle\langle011| \tag{8.71}$$

$$P_2 = |010\rangle\langle010| + |101\rangle\langle101| \tag{8.72}$$

$$P_3 = |001\rangle\langle001| + |110\rangle\langle110| \tag{8.73}$$

where P_0 detects no error, P_1 detects a first bit flip, P_2 detects a second bit flip, and P_3 detects a third bit flip. These measurements do not collapses the state of the code.

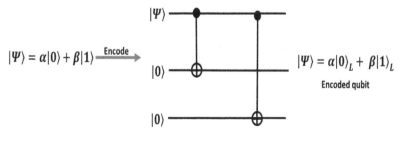

Circuit for encoding a qubit

FIGURE 8.7 An illustration of a three-qubit encoding circuit.

Consider that after some time period, a first bit flip has occurred, and the state of the encoded qubit is:

$$|\Psi\rangle_{\text{error}} = c_0|100\rangle + c_1|011\rangle \qquad (8.74)$$

All the measurement operators commute, and so the measurements can be carried out simultaneously. By performing all projection operators simultaneously, the error can be detected. The outcome of all measurements yields a 0 value except for P_1. The outcome of such a measurement is:

$$\langle\psi_{\text{error}}|P_1|\psi_{\text{error}}\rangle = 1 \qquad (8.75)$$

Such a measurement detects the error. The state of the encoded qubit remains the same after measurement:

$$|\Psi'\rangle = \frac{P_1|\Psi\rangle_{\text{error}}}{\sqrt{\langle\Psi_{\text{error}}|P_1^\dagger P_1|\Psi_{\text{error}}\rangle}} = |\Psi\rangle_{\text{error}} \qquad (8.76)$$

After determining the error in the state of the qubit, error-correction is performed. By simply flipping the state of the first qubit, the corrected state of the qubit is obtained with perfect accuracy. This is an interesting approach because errors for a quantum state can be detected and corrected without destroying the state of the qubit. The measurement collapses the quantum superposition state, however, for quantum error-correcting codes, measurement helps to detect the error.

PROBLEMS

8.6 Consider the following state of a three-qubit code that has undergone an error, $|\Psi\rangle_{\text{error}} = c_0|101\rangle + c_1|010\rangle$. Show that such an error cannot be corrected by using one of the above projection operators.

8.7 Find the projection operator that can detect error in the following state: $|\Psi\rangle_{\text{error}} = c_0|001\rangle + c_1|110\rangle$.

8.8 Prove that only single bit-flip errors can be detected by applying the projection operators above. Correlated errors cannot be detected.

8.8   QUANTUM TELEPORTATION

The word teleportation is commonly used in science fiction, where it means transport of energy or matter from one location to another location without traversing space. Quantum teleportation is not the transfer of energy or matter. But rather the transfer of a quantum state from one location to another using classical communication and entanglement.

Consider the case of Alice wishing to transport a quantum state to Bob. Bob and Alice are physically separated in space. To do this, she and Bob share a two-qubit entangled state expressed as:

$$|\varphi\rangle = \frac{1}{\sqrt{2}}|00\rangle + \frac{1}{\sqrt{2}}|11\rangle \qquad (8.77)$$

Alice wants to transport the following state to Bob without physically sending it to Bob:

$$|\Psi\rangle = c_0|0\rangle + c_1|1\rangle \qquad (8.78)$$

The resulting state is:

$$|\Psi'\rangle = (c_0|0\rangle + c_1|1\rangle)\left(\frac{1}{\sqrt{2}}|00\rangle + \frac{1}{\sqrt{2}}|11\rangle\right) \tag{8.79}$$

In the equation above, the first qubit is the one Alice wants to teleport to Bob, and the second and the third qubits are the shared entangled pairs. Alice sends her two qubits through a C-NOT gate, and the resulting state is obtained:

$$|\Psi'\rangle = c_0|0\rangle\left(\frac{1}{\sqrt{2}}|00\rangle + \frac{1}{\sqrt{2}}|11\rangle\right) + c_1|1\rangle\left(\frac{1}{\sqrt{2}}|10\rangle + \frac{1}{\sqrt{2}}|01\rangle\right) \tag{8.80}$$

Now, Alice sends the first qubit through a Hadamard gate operation, and she obtains the following state:

$$|\Psi'\rangle = \frac{1}{2}(|00\rangle(c_0|0\rangle + c_1|1\rangle) + |01\rangle(c_0|0\rangle + c_1|1\rangle) + |10\rangle(c_0|0\rangle - c_1|1\rangle) + |11\rangle(c_0|0\rangle - c_1|1\rangle)) \tag{8.81}$$

The first two qubits in the above equation are in Alice's possession. By performing measurements on two qubits she can teleport a required state to Bob.

Let us say Alice performs a measurement on the first two qubits, and found them to be in state, $|00\rangle$, the state of the qubit that Bob has now is:

$$|\Psi\rangle = c_0|0\rangle + c_1|1\rangle$$

Thus, she has successfully teleported the required state to Bob. All she has to do is to call Bob, and tell him to do nothing to the qubit in his possession.

If Alice performs the measurement on the two qubits and determines that they are in state $|01\rangle$, she can perform the same operation as above.

However, if she finds the two qubits in her possession to be in state $|10\rangle$ or $|11\rangle$, she discovers the state of the qubit in Bob's possession. She calls Bob and tells him the state of the qubit in his possession. He then performs the right gate operation (Z or X gate), and obtains the state of his qubit as:

$$|\Psi\rangle = c_0|0\rangle + c_1|1\rangle$$

Thus, the state of the qubit is teleported without being transported through space. Is the quantum state teleported faster than the speed of light? The answer is No because to teleport such a state, Alice must communicate with Bob over a classical communication channel. Without this classical channel, Alice cannot convey any information to Bob. Any measurement performed by Bob on the qubit in his possession will lead to teleportation. This demonstrates further the power of quantum phenomenon such as entanglement. What is really teleported here? Is it information?

To answer this question, consider a scenario, Alice has a secret message that she needs to communicate to Bob, without sending the message physically. Such a message can be teleported utilizing the above technique. For this, Alice and Bob share multiple qubits as entangled pairs. By Alice performing the same operations as discussed in the previous paragraph she covert the states of the qubits in Bob's possession into a code which Bob can decipher. In this way Bob can receive the message from Alice without the need for Alice to communicate anything to Bob through a communication channel. Thus, information is transferred to Bob without the need for physical transport of qubits. Since, there is no transfer of the energy in the process, this transfer of information does not violate relativity principle of faster than the speed of light.

8.9   SUPERDENSE CODING

Superdense coding is an example of the use of quantum properties where quantum information can be transmitted in a way that classical information cannot. The idea is to transmit a maximum amount of information by employing a minimum number of qubits.

Consider that Alice and Bob share an entangled pair as specified below:

$$|\Psi\rangle = \frac{1}{\sqrt{2}}|00\rangle + \frac{1}{\sqrt{2}}|11\rangle \tag{8.82}$$

Initially, the first qubit is in the possession of Alice and the second qubit is in the possession of Bob. The state of Equation (8.82) is an entangled state. Now, Alice wants to transmit two bits of information to Bob by sending a single qubit. If she wished to transmit bit "00" to Bob, she does nothing to her qubit, and just transmits it to bob. If she wishes to transmit bit "01" to Bob, she performs a phase flip, Z gate operation to her qubit and then transmits it to Bob. If she wishes to transmit bit "10" to Bob, she performs a NOT gate, X gate operation to her qubit and then transmits it. If she wishes to transmit bit "11" to Bob, she performs an iY gate operation to her qubit and then transmits it. The resulting states are:

$$|\Psi_{00}\rangle = \frac{1}{\sqrt{2}}|00\rangle + \frac{1}{\sqrt{2}}|11\rangle \tag{8.83}$$

$$|\Psi_{01}\rangle = \frac{1}{\sqrt{2}}|00\rangle - \frac{1}{\sqrt{2}}|11\rangle \tag{8.84}$$

$$|\Psi_{10}\rangle = \frac{1}{\sqrt{2}}|10\rangle + \frac{1}{\sqrt{2}}|01\rangle \tag{8.85}$$

$$|\Psi_{11}\rangle = \frac{1}{\sqrt{2}}|01\rangle - \frac{1}{\sqrt{2}}|10\rangle \tag{8.86}$$

These four states are known as Bell states or a Bell basis. All the Bell states are entangled and forms an orthonormal basis. Thus, these states can be distinguished by performing a measurement. Since Alice has transmitted her qubit to Bob who is now in possession of both qubits, by performing measurements on a Bell basis, Bob can determine which of the four possible two bits were sent by Alice. Hence, Alice was able to transmit two bits of information just by transmitting a single qubit to Bob. Alice interacts only with the qubit that is in her possession and was able to send two bits of information to Bob. Classically, Alice would need to interact with Bob's qubit for transmitting two bits of information. Such a superdense transmission of information is only possible in a quantum computer.

Here we end this chapter without any final conclusions about the field of quantum information and quantum computation. The main reason for doing so is that the field is evolving very rapidly. It is hard to be clear about the future technologies emerging from this field. But it is clear that future quantum technologies will be built on the principles that are very close to the principles that govern this universe.

References

BOOKS

1. D'Abro, A. 1951. *The Rise of the New Physics, Its Mathematical and Physical Theories*. New York: Dover Publications Inc. Vol. II.
2. Bohm, D. 1951. *Quantum Theory*. New York: Dover Publications Inc.
3. Dirac, P.A.M. 1967. *The Principles of Quantum Mechanics*. Oxford: Clarendon Press. Fourth edition.
4. Heisenberg, W. 1949. *The Physical Principles of the Quantum Theory*. New York: Dover Publications Inc.
5. Feynman, R., Leighton, R., & Sands, M. 1966. *The Feynman Lectures on Physics*, Vol. III: Quantum Mechanics. New York: Basic Books.
6. Sakurai, J.J. 1999. *Modern Quantum Mechanics*. Reading, MA: Addison-Wesley Publishing Company Inc.
7. Cohen Tannoudji, C., Diu, B., & Laloe, F. 2005. *Quantum Mechanics*. Toronto, Canada: John Wiley & Sons, Pte, Ltd. Vol. I.
8. Shankar, R. 1994. *Principles of Quantum Mechanics*. New York: Kluwer Academic/Plenum Publishers. Second edition.
9. Schiff, L.I. 1968. *Quantum Mechanics*. New York: McGraw-Hill. Third edition.
10. Griffiths, D.J. 2005. *Introduction to Quantum Mechanics*. Hoboken, NJ: Pearson Education Inc. Second edition.
11. Nielsen, M.A. & Chuang, I.L. 2000. *Quantum Computation and Quantum Information*. New York: Cambridge University Press.
12. Byron, F.W. Jr. & Fuller, R.W. 1992. *Mathematics of Classical and Quantum Physics*. New York: Dover Publications Inc.
13. Landau, L.D. & Lifshitz, E.M. 1976. *Mechanics*. Oxford: Pergamon Press. Vol. I, Third edition.
14. McDermott, L.C. 1996. *Physics by Inquiry*. Hoboken, NJ: John Wiley & Sons, Inc. Vol. II.
15. Heisenberg, W. Reprinted in 1990. *Physics & Philosophy, the Revolution in Modern Science*. London: Penguin Group.

JOURNAL PAPERS

16. Kramm, G. & Molders, N. 2009. Planck's blackbody radiation law: Presentation in different domains and determination of the related dimensional constants, *arXiv preprint arXiv:0901.1863*.
17. Davisson, C. & Germer, L.H. Diffraction of electrons by a crystal of nickel. *The Physical Review*, 30(6), 705.
18. Gehrenbeck, R.K. 1978. Electron diffraction: Fifty years ago. *Physics Today*, 31(1), 34–41.
19. Stodolna, A.S., Rouze, A., Lépine, F., Cohen, S., Robicheaux, F., Gijsbertsen, A., Jungmann, J.H., Bordas, C. & Vrakking, M.J.J. 2013. Hydrogen atoms under magnification: Direct observation of the nodal structure of stark states. *Physical Review Letters*, 110(21), 213001.
20. Hansch, T.W., Schawlow, A.L. & Series, G.W. 1979. The spectrum of atomic hydrogen. *Scientific American*, 240(3).
21. Zurek, W.H. 2002. Decoherence and the transition from quantum to classical-revisited. *Los Alamos Science*, 27, 86–109.
22. Hornberger, K. 2009. Introduction to decoherence theory. *Lecture Notes in Physics*, 768. Springer-Verlag Berlin Heidelberg.
23. Aaronson, S. 2008. The limits of quantum computers. *Scientific American Inc.*, 298(3), 62–69.
24. Mosca, M. 2008. Quantum algorithms. *arXiv:0808.0369v1 [quant-ph]* 4.
25. Siddiqui, S. & Singh, C. 2017. How diverse are physics instructors' attitudes and approaches to teaching undergraduate level quantum mechanics? *European Journal of Physics*, 38(3).

References

BOOKS

Index

A

Absorption spectrum, 153
Adiabatic approximation, 196
Adiabatic computation, 226
Adiabatic invariant in pendulum, 197
Adiabatic process, 197
Adiabatic theorem, 197–199
  proof of, 199–200
Akira Tonomura, 27
Algebraic method, 134–137
Algorithm, 209
Alice and Bob, 219, 230
Alpha decay theory, 99
Alpha particle, 99, 184
Analytical method, 127–131
AND logic gates, 209
Angular equation, 142
Angular momentum, 155–158, 162–165
Angular momentum quantum number, 143
Anti-Hermitian, 45
Associated Laguerre polynomials, 151–152
Associated Legendre function, 144
Asymptotic behavior, 147–148
Average angular momentum, 160
Average energy, 8, 81, 85, 129
Azimuthal angle, 141
Azimuthal quantum number, 143

B

Balmer series, 155
Balmer, Johann Jakob, 154
Base kets, 44–45
Bell, J. S., 206
Bell's inequality, 206–207
Berry, M. V., 198, 215
Berry's phase, 199
Beta decay, 195–196
Bits, 209
Black body, 4–5
Black body paradox, 7
Black-body radiation paradox, 4
  experimentalobservations, 5–6
  radiation function, mathematical form of, 7–9
Bloch sphere, 73, 213
Bohr, Niels, 1
Bohr radius, 150
Born, Max, 197
Born's statistical interpretation, 55
Bound state, 93
Boundary conditions, 63, 75, 140–141, 186
Bra space, 42; *see also* Dual space
Bra vectors/bras, 42
Bragg's condition, 26

B (continued)

Bragg's law, 26
Bytes, 209

C

Case I (E>V0), 93–94
Case II (E<V0), 94–97
Center of mass, 146–147
Central processing unit (CPU), 209
Centrifugal force, 145
Characteristic frequency of vibration, 7–8
Classical and quantum Harmonic oscillator, comparison, 132–133
Classical and Quantum mechanical particle, comparison, 85
Classical and quantum mechanical view of radiation, comparison, 10
Classical computer, 209–211
Classical mechanics, 39
  approach, 39
Classical theory of electromagnetism, 11
Claude, George, 154
Closed loop, 199
Closure property, 47–48
C-NOT gate operation, 217
Collapse of the state, 48–49
Commutator bracket, 65
Compatible observables, 66
Completeness relation/closure, 45
Complex valued function, 51
Complex vector space, 41, 47, 67
Compton effect, 14–15
Compton recoil, 68
Compton, Arthur Holly, 1, 14
Computational basis, 214–215
Condition of continuity, 94, 102
Condon, 184
Conservation of probability, 64–65
Conserved quantity, 64–65, 157–158, 160
Constructive interference, regions of, 24
Continuous spectrum, 50
Continuous variable, 42, 50–52
Copenhagen interpretation, 37–38, 203, 225
Coulomb's law, 147
Coulombic force, 184
Coulombic potential, 153, 185
CPU, *see* Central Processing Unit (CPU)

D

Davisson, Clinton, 25
De Broglie matter waves, 21–25
De Broglie wave, 82, 181
De Broglie, Louis, 21

De Broglie's momentum relationship, 15
De-Broglie wave, velocity, 22
De-Broglie's momentum-wave relationship, 22
Decoherence theory, 225
Decoherence, 224–226
    Time, 225–227
Decoherence-free subsystem/subspace, 226
Degeneracy of Hydrogen atom, 153
Degenerate eigenvalues, 49
Degenerate state, 170–172
Degenerate perturbation, 170–172
    degenerate state, 170–172
Density operator, 222
    properties, 223–224
    reduced density operator, 224
Destructive interference, regions of, 24
Deutsch, David, 218
Deutsch's algorithm, 218–219
Deutsch's-Jozsa algorithm, 219–220
Differential radiant energy, 6
Diffraction, 1, 26
Dimensions, 71
Diodes, 2
Dirac, P. A. M, 134
Dirac delta function, 51
Discontinuous process, 8
Discrete band, 154
Discrete energy, 1, 21, 88
Discrete nature of light, 1
Discrete spectrum, 48–50
Discrete variable, 42
Discretization of energy, 88
Dispersion, 12, 120, 124
Double-slit experiment, 27–32
Driving frequency, 193
Dual space, 42
Dynamic phase factor, 198
Dynamical properties, 59
    separation of variables, 61–62
    stationary states, 62–63
    time-dependent schrodinger equation,
        59–60
    time evolution operator, 60–61
Dynamical variables, 39

E

Ehrenfest, Paul, 197
Eigenfunctions, 50, 53–54
    completeness, 54
    normal, 54
    orthogonal, 53
Eigenstates, linear superposition of, 41
Eigenvalue, 48
Eigenvector of operator, 72
Einstein, Albert, 1, 197, 202
Einstein, Podolsky and Rosen (EPR), 204
Einstein's interpretation, 206
Electric fields, 2
Electrical discharge tube, 155

Electromagnetic wave, 1–2
    generated, 3–4
    properties, 3
        frequency in a vacuum, 3
Electron
    diffraction, 25–27
    gun, 30
    magnetic moment of, 39–40
    electron microscope, 2
    shield, 177
    spin/intrinsic angular momentum, 40
    wave function, 30
Electrostatic attraction, 176
Electrostatic repulsion, 176
Empirical law, 6
Energy eigenfunction, 52
Energy eigenstates, 63
Energy quantization rule, 187
Energy-time uncertainty relationship, 66
Entangled state, 203, 215, 217, 228
EPR, *see* Einstein, Podolsky and Rosen (EPR)
EPR paradox, 204–206
Equilibrium position, 132
Equilibrium radiation, 5
Error-correction, 226
Euclidean space, 46
Euclidean space and vector space, comparison, 47
Everett III, Hugh, 225
Expansion coefficient, 52, 61, 90, 106–107, 151,
        191–193
Expectation value, 56, 91, 129, 160, 174, 176, 178
Experimental value, 179
Exponentially decaying function in Region II, 96

F

Feynman, Richard, 34
Fine structure of the spectrum, 155
Finite square well, 100–103
First application of quantum mechanics, 184
First few spherical harmonics, 145
First three stationary states, 89
First-order, 190–193
First-order corrections, 168–169
First-order perturbation correction, 168
First-order perturbation theory, 190–193
Floating Point Operations Per Second (FLOPS), 209
FLOPS, *see* Floating Point Operations Per Second
        (FLOPS)
Fok, Vladimir A., 197
Forbidden region, 186
Formalism, 39–79
    basic postulates of the model, 39
        indeterminacy, 40
        postulate (1), 39–40
        postulate (2), 40–41
        postulate (3), 41
    conservation of probability, 64–65
    dynamical properties, 59
        separation of variables, 61–62

stationary states, 62–63
time-dependent schrodinger equation, 59–60
time evolution operator, 60–61
Heisenberg uncertainty principle, 65–67
mathematical foundation, 41
basekets, 45
completeness, 45
continuous spectrum, 50–51
discrete spectrum, 48–50
ket and bra, multiplication rules for, 43
inner product, 43
normalization, 43–44
outer product, 43
orthogonality, 43
ket and bra space, 42
operators, 45
state vector, 41–42
matrix mechanics, 71–72
tutorials, 73
spin (1/2), 73–79
wave function, 51
eigenfunctions, 53
orthogonal, 53
normal, 54
completeness, 54
expectation value of an observable, 56
position, 56
momentum, 57
kinetic energy, 57–58
function space, 51–53
hilbert space, 51–53
probability density, 54–56
uncertainty principle, proof of, 67
principle applications, 68
Heisenberg's microscope, 68–69
orbits in atoms, 69–70
Schwarz inequality, the, 67–68
Formalism of quantum mechanics, structure, 57
Formalism-I applications, 81
finite square well, 100–103
free particle, 81–86
infinite square well, 86–92
potential barrier penetration (tunneling), 98–100
quantum wave packet, 119–120
step potential, 92–98
tunneling, 110–111
tutorials, 105
infinite square well, 105
Formalism-II applications, 127
angular momentum, 155–158
eigenfunctions and eigenvalues, 159
harmonic oscillator, 127
algebraic method, 134–137
normalization constant, 137–138
analytical method, 127–131
normalization constant, 131–133
hydrogen atom, 146
spectrum, 153–155
radial equation, 147
asymptotic behavior, 147–148
laguerre polynomials and associated, 151–152

hydrogen atom, degeneracy of, 153
power series solution, 148–151
Schrodinger equation in three dimensions, 138
angular equation, the, 142–145
in Cartesian coordinates, 138–140
new quantum property, 140–141
radial equation, 145
in spherical coordinates, 141–142
tutorial, 162
angular momentum, 162–165
Free electrons, 14–15, 37
Free particle, 81–86
Function space, 51–53

G

Gamma waves, 4
Gamow, George, 184
Gamow's model, 185
Gaussian wave packet, 83–84
Geometric phase, 199, 202
Geometric properties of the system, 199
Germer, Lester, 25
Giorgio, Pier, 27
Global property, 218–219
Ground state energy eigenfunction, 130
Ground state of helium atom, 179
Ground state wave function, 150, 174, 177
Group velocity, 24, 85
Gurney, 184

H

Hamiltonian, 52
Operator, 52, 59–61, 64, 81, 134, 136, 143, 159
Harmonic oscillator, 127–138
algebraic method, 134–137
normalization constant, 137–138
analytical method, 127–131
normalization constant, 131–133
Schrodinger equation, 128
Haocheng Yin, 24
Heinrich Hertz, 11
Heisenberg uncertainty principle, 65–67
Heisenberg's microscope, 68–69
Heisenberg's microscope, 69
Heisenberg's uncertainty principle, 65
Heisenberg, Werner, 1
Helium atom, 176–177
Hermite polynomial, 130–131
Hilbert space, 51–53
Huggins, Sir William, 154
Hydrogen atom spectrum, 153–155
Hydrogen gas, 153–154
Hydrogen atom, 146
hydrogen atom spectrum, 153–155
radial equation, 147
asymptotic behavior, 147–148
hydrogen atom, degeneracy of, 153
Laguerre polynomials and associated, 151–152

Hydrogen atom (*Continued*)
  power series solution, 148–151
Hydrogen spectral tube, 155

I

Incident wave, 93
Incompatible observables, 66
Infinite square well, 86–92, 105
 potential, 87
Inherent temperature, 4
Inner product, 43
Interference, 1
Interference of waves, 28
Interference term, 30
Intrinsic angular momentum, 39

J

Jonsson, Clauss, 27

K

Ket space, 42
Kets, 40
Kinetic energy, 11
 operator, 57–58
Kirchhoff's law, 5–6
Known state, 213
Kronecker delta function, 54

L

Laguerre polynomial, 151
Laplacian, 138, 141
Larmor frequency, 73
Lasers, 2
Legendre function, 144
Legendre polynomial, 144
Linear Hermitian operator, 40
Linear superposition of
   eigenstates, 41
Logical gate, 209
Lowering operator, 135–136
Lowest energy, 129–130, 136
Lyman series, 155

M

Magnetic fields, 2
Magnetic moment, 39–40, 73
Magnetic quantum number, 144
Magnetic resonance imaging, 2
Many worlds Interpretation, 225
Matrix element, 71
Matrix mechanics, 1, 71–72
Maxima, 28
Maxwell's theory of light, 1
Measurement, 36–38

Memory unit, 209
Metal electrodes, 11
Metal plate, 11–12, 17, 37
Microwaves, 4
Minima, 28
Missiroli, GianFranco, 27
Mixed state, 222
Momentum operator, 57
Monochromatic energy density, 8
Monochromatic radiation, intensity of, 5
Moore's law, 210
Multiple qubits, 214
 entanglement, 215

N

Nickel crystal, planar surface of, 26
Nobel prize, 12, 27
Nodes, 89
Nondegenerate eigenvalues, 49
Non-degenerate perturbation, 168
 first-order correction, 168–169
 second-order correction, 170
Normalization, 43–44
 constant, 131–133, 137–138
NOT logic gates, 209
Novel theory, 1

O

Odd parity, 172, 174
One-dimensional harmonic oscillator, 192
Operators, 45
Opposite parity, 173
Orthogonality, 43
Orthonormal base kets, 46
Orthonormality condition, 54, 195
Oscillating charged particle, 10
Oscillating wave function in Region I, 96
Oscillator, energy density function of, 8
Outer product, 43
Overcoming decoherence, methods for, 226
 quantum error-correcting codes, 226–227

P

Particle in a box, 105
Paschen series, 155
Periodic perturbations, 193–194
Perturbation theory, 167
Perturbation theory, 167–170
 time-independent perturbation, 167–170
  degenerate perturbation, 170–172
   degenerate state, 170–172
  non-degenerate perturbation, 168
   first-order correction, 168–169
   second-order correction, 170
  Stark effect, 172–173
   parity, 172–173
  variational principle, 174–176

ground state of a helium atom, 176–180
WKB method, 180–186
turning points of a bound state, 186–188
Perturbed energies, 173
Perturbed system, 167–168, 192
Phase velocity, 84
Photocurrent, 12
Photoelectric effect, 11–14, 17–20
Photoelectrons, 12
Photon wave packet, 86
Photons, 9
Planck, Max, 1
Planck's constant, 8, 73
Planck's radiation law, 9
Planck's radiation model, 7–8
Podolsky, Boris, 204
Polar angle, 141, 178
Polished silver, 4
Polynomial time, 210
Potential barrier penetration (tunneling), 98–100
Region I, 98
Region II, 98
Region III, 99–100
Potential energy function, 127
Power series solution, 148–151
Principal quantum,153
Probability amplitude, 31
Probability density, 31, 50, 54–56
plots, 132
Probability function, 54
Probability waves, 32, 87
Projection measurement operators, 227
Pure state, 222
Puzzi, Giulio, 27

Q

Quanta of energy, 9, 37
Quanta, 1, 9
Quantization of energy, Planck's idea, 1
Quantum algorithms, 218–220
Quantum bit, 212
Quantum computer, 209, 211
classical computer, 209–211
decoherence, 224–226
density operator, 222
properties, 223–224
reduced density operator, 224
methods for overcoming decoherence, 226
quantum error-correcting codes, 226–227
quantum algorithms, 218–220
quantum computer, 211
multiple qubits, 214
entanglement, 215
qubit, 211–214
qubit gates, 215–217
quantum measurement, 220–222
quantum teleportation, 228–230
superdense coding, 230
Quantum effect, 154, 211

Quantum error correcting code, 226
Quantum error-correction, 226
Quantum gates, 215–216
Quantum jumps, 189
Quantum measurement, 220–222
Quantum mechanical approach, 39
Quantum mechanical system, 39
defining an observable, 39–40
indeterminacy, 40
measurement of an observable, 41
states associated with an observable, 40–41
Quantum mechanics-I, history
beginning view, 9–10
black-body radiation paradox, 4
experimentalobservations, 5–6
radiation function, mathematical form
7of, 7–9
Compton effect, 14–16
electromagnetic wave, classical view of, 1
generated, 3–4
properties, 3
mathematical theory, 1
photoelectric effect, 11–14
tutorial, 17
photoelectric effect, 17
Quantum mechanics-II, history
De Broglie matter waves, 21–25
double-slit experiment, 27–32
electron diffraction, 25–27
measurement, 36–38
uncertainty principle, 32
simple proof of, 35
Quantum numbers, 150
Quantum teleportation, 228–230
Quantum theory, 8
Quantum tunneling, 113
Quantum wave packet, 119–120
Quantum-mechanical constants, 143
Qubit, 211–214
gates, 215–217

R

Radial equation, 142, 145
Radial function, 150–151
Radiation, 4
Radiation gas, 5
Radiation, intensity function of, 8
Radio waves, 4
Raising operator, 135–136
Rayleigh–Jeans law, the black-body
paradox, 7
Rectangular potential barrier, 110
Recursion formula, 149, 151
Reduced density operator, 224
Reflected wave, 93
Reflection coefficient, 99
Reflection property, 94
Relative potential energy, 146
Relativistic electron, 195

Repulsive potential, 145
Ripples in a pond as waves, 2
Rodrigues formula, 144
Rosen, Nathan, 204

S

Scattered photon, 33
Scattering state, 93
Schrodinger, Erwin, 1
Schrodinger equation in three
          dimensions, 138
    angular equation, the, 142–145
    in Cartesian coordinates, 138–140
        new quantum property, 140–141
    radial equation, 145
    in spherical coordinates, 141–142
Schwarz inequality, the, 67–68
Second-order correction, 170
Semi-classical approximation, 180
Separation of variables, 61–62
Shallow well, 103
Shor, P., 226
Shor's algorithm, 220
Sinusoidal function, 181
Skew Hermitian, 45
Soot, 4–5
Spectral energy density, 8
Spectrum of continuous band, 154
Spherical harmonics, 144
Spherical symmetry, 141
Spin (1/2), 73–79
Spin angular momentum, 155, 205–206
Spin operator, eigenkets of, 42
Spin singlet state, 202
Spin/intrinsic angular momentum, 40
Spin-down orientations, 40
Spinor, 71
Spin-up ket vector, 71
Spin-up orientations, 40
Splitting of the energy, 172
Standard deviation, 34
Stark effect, 172–173
    parity, 172–173
State of statistical equilibrium, 5
State vector, 41–42, 72
    collapse, 49
States, 40
Stationary electron, 15–16
Stationary states, 62–63
Stationary state wave function, 82
Steane, A. M, 226
Stefan-Boltzmann's law, 6
Stellar spectra, 154
Step potential function (barrier), 95
Step potential, 92–98
Sudden approximation, 194–196
Superdense coding, 212, 230
Superposition of states, 40
Superposition principle, 211–212

T

Telecommunication, 2, 209
Theory of relativity, 22
3-Dimensional infinite square well potential
    degenerate energy levels of, 140
    visualization of, 140
3-Dimensional Schrodinger equation, 138
Three-qubit encoding circuit, 227
Time dependent perturbation
Time evolution operator, 60–61
Time-dependent perturbation theory, 189
    adiabatic approximation, 196
        adiabatic process, 197
        adiabatic theorem, 197–199
            proof of, 199–200
    first-order, 190–193
    measurement problem revisited, 202
        Bell's in equality, 206–207
        EPR paradox, 204–206
    periodic perturbations, 193–194
    sudden approximation, 194–196
Time-dependent potentials, 189
Time-dependent schrodinger equation, 59–60
Time-dependent stationary state, 90
Time-independent perturbation, 167–170
    degenerate perturbation, 170–172
        degenerate state, 170–172
    non-degenerate perturbation, 168
        first-order correction, 168–169
        second-order correction, 170
    Stark effect, 172–173
        parity, 172–173
Time-independent potential function, 189
Time-independent Schrodinger equation, 62, 87, 141, 180
Toffolli gate, 216
Transcendental equation, 102
Transistors, 2, 211
Transition frequency, 193
Transition probability, 194
Transition region, 187
Transmission coefficient, 99
Transmitted wave, 93–94, 97, 99
Tunneling, 110–111
Tunneling through a barrier, 184–185
Turning point, 93
Tutorials, 105
    infinite square well, 105
Two entangled particles, measurement outcomes of, 203
Two-level system, 191

U

Ultraviolet catastrophe, 7
Ultraviolet light, 11
Uncertainty principle, 32
    simple proof of, 35
Uncertainty principle, proof of, 67
    principle applications, 68
        Heisenberg's microscope, 68–69

orbits in atoms, 69–70
Schwarz inequality, the, 67–68
Unit vectors, 47
Unitary operations, 216
Unknown parameter, 177

V

Variational principle, 174–176
ground state of a helium atom, 176–180
Vectors, 40
Vector space, 40–44, 46–47, 51, 58, 67, 71
Visualization tool, 73, 214

W

Wave crests, 22
Wave equation, 82
Wave function, 51–58
eigenfunctions, 53
completeness, 54
normal, 54
orthogonal, 53
expectation value of an observable, 56
kinetic energy, 57–58
momentum, 57
position, 56
function space, 51–53
Hilbert space, 51–53

probability density, 54–56
Wave mechanics, 1, 22
Wave packet, 23–24, 83
Wavelength dependent intensity function, 8
Wave-particle duality, analogy of, 37
Wein's displacement law, 6
Wein's relation, 7
Wentzel-Kramers-Brillouin (WKB), 180
WKB, *see* Wentzel-Kramers-Brillouin (WKB)
WKB approximation, 181
WKB method, 180–186
turning points of a bound state, 186–188
Words, 209

X

X-ray photons, 37
X-rays, 4

Y

Young, Thomas, 27

Z

Z (phase gate), 216
Zeh, H.D., 225
Zurek, W.D., 225

Printed and bound by CPI Group (UK) Ltd, Croydon, CR0 4YY

17/10/2024

01775694-0010